LASERS AND ELECTRO-OPTICS RESEARCH AND TECHNOLOGY

NONLINEAR OPTICS

FUNDAMENTALS, APPLICATIONS AND TECHNOLOGICAL ADVANCES

LASERS AND ELECTRO-OPTICS RESEARCH AND TECHNOLOGY

Additional books in this series can be found on Nova's website under the Series tab.

Additional e-books in this series can be found on Nova's website under the e-book tab.

LASERS AND ELECTRO-OPTICS RESEARCH AND TECHNOLOGY

NONLINEAR OPTICS

FUNDAMENTALS, APPLICATIONS AND TECHNOLOGICAL ADVANCES

FRED WILKINS
EDITOR

nova publishers
New York

For permission to use material from this book please contact us:
Telephone 631-231-7269; Fax 631-231-8175
Web Site: http://www.novapublishers.com

Library of Congress Cataloging-in-Publication Data

Nonlinear optics : fundamentals, applications and technological advances / [edited by] Fred Wilkins.
pages cm. -- (Lasers and electro-optics research and technology)
Includes bibliographical references and index.
ISBN 978-1-63321-926-7 (hardcover)
1. Nonlinear optics. I. Wilkins, Fred, 1969- editor.
QC446.2.N676 2014
621.36'94--dc23

2014033259

Published by Nova Science Publishers, Inc. † New York

CONTENTS

PREFACE

In recent years, there is increasing interest in the research on optical fiber nanowires or microfibers (MFs) which are fibers with submicrometer- and nanometer- diameter. The size is tens to thousands of times thinner than the standard optical fiber. With the advantages of large evanescent fields, high nonlinearity, extreme flexibility and configurability and low-loss interconnection to other optical fibers and fiberized components, they have found potential applications in a wide range of fields of optical communications, sensing, lasers, biology and chemistry. This book discusses the fundamentals, applications and technological advances of nonlinear optics.

Chapter 1 - Diamond-like semiconductors (DLSs), which are inherently noncentrosymmetric, exhibit a wide range of compositional flexibility and thus comprise an attractive platform for physical property tuning for an array of potential nonlinear optical (NLO) applications. Although ternary DLSs such as $AgGaS_2$, $AgGaSe_2$ and $ZnGeP_2$ have long dominated the list of benchmark NLO materials, only recently have investigations into the NLO properties been initiated for quaternary DLSs, which offer a higher degree of versatility. The authors have studied the efficiencies of second harmonic generation (SHG) and third harmonic generation (THG) as a function of broadband wavelength for several relatively unexplored quaternary DLSs. Cu_2CdSnS_4 adopts the stannite structure that is based on the cubic diamond structure, crystallizing in space group I-*42m*. α- and β-Cu_2ZnSiS_4 exist as two diamond-like polymorphs crystallizing in the wurtz-stannite (P*mn2*$_1$) and wurtz-kesterite (P*n*) structures, respectively, which are derived from the hexagonal lonsdaleite structure. Li_2CdGeS_4 forms with the wurtz-stannite structure. These materials exhibit relatively high thermal stability and are air stable. The selection of elements used to construct these materials plays a key role in the resulting physicochemical properties such as bandgap energy. Electronic structure calculations provide insight into how the composition affects the magnitude of the bandgap, which in turn influences the optical transparency window and laser-damage threshold (LDT). The energy gaps of these semiconductors range from ~ 0.9 to 3.2 eV. Typically, enhanced second-order nonlinearity has been expected for relatively narrow-gap materials, but highly efficient generation of long wavelengths requires a sufficiently wide bandgap. The narrow-gap semiconductor Cu_2CdSnS_4 exhibits high SHG and THG responses in the IR, but practical applications are limited by LDT, significant one-photon absorption at 1064 nm. On the other hand, the relatively wide-gap semiconductors α- and β-Cu_2ZnSiS_4 exhibit a high LDT and a wide transparency window, but relatively low SHG and THG responses. Li_2CdGeS_4, an especially unique material, possesses a wide bandgap and exhibits high SHG and THG efficiencies along with an exceptional LDT, demonstrating that high coefficients of nonlinearity and wide bandgaps are not mutually exclusive.

Chapter 2 - Alkali atoms have large electric dipole moments due to their unique optically active electron in the external shell. Therefore, the coupling between an electric field and the electron is strong, making alkali vapors very suitable media for nonlinear optics studies and applications. Indeed, the observation of nonlinear behavior around atomic resonances is possible with low power of continuous laser light. Moreover, as these resonances are narrow compared to the tuning range of most lasers, one can explore different signs and strengths of the atom-light interaction by only slightly changing the laser frequency. One of the simplest nonlinear phenomenon in alkali atoms originates from a Kerr-like term in the refractive index, an effect involving the lowest nonlinear susceptibility for an center-symmetric media. In the regime where the Kerr effect becomes relevant, this interaction changes the refractive index of the medium, which results in a modification of the beam propagation in a self-interaction process. For a laser beam with a radial gradient of intensity (as in the typical case of Gaussian beams) a radial gradient of refractive index will appear in such a way that the medium will exhibit a lens-like behavior. This phenomenon, known as self-lensing, is relatively easy to be implemented experimentally, allowing the measurement of the Kerr coefficient of resonant atomic vapors, which appears as an interesting application of the z-scan technique. Indeed, this technique has been extensively used for measurements in liquids and solids, but relatively few experiments have applied it to thermal atomic vapors or cold atoms clouds. In this chapter, the authors first review the interaction between a laser beam and alkali atoms in order to characterize the origin of the nonlinear refractive index and to explore the rich dependency of such a phenomenon with the frequency detuning around the sharp resonance. Then the authors discuss the z-scan technique applied to cold atoms clouds and atomic vapors, as well as recent experimental investigations.

Chapter 3 - Nowadays, two-photon excitation and other nonlinear optical approaches have been successfully used within the microscopy framework which evolved as an alternative to conventional single-photon confocal microscopy. In general, they are collectively known as nonlinear microscopy, which includes two-photon excited fluorescence (TPEF) microscopy, second, and third harmonic generation (SHG, THG) microscopy, and coherent anti-Stokes Raman scattering (CARS) microscopy. TPEF approach provides high-resolution images of molecular autofluorescence signals in living tissues, SHG microscopy has been successfully used to study structural protein arrays, such as collagen, while CARS and THG signal can be used both to reveal the presence of lipid bodies and to highlight the nuclei inside the thickness of a tissue.

In the first section of this chapter will be described the nonlinear contrast mechanism of each technique, and how to combine and build these setup, on a single commercial multimodal confocal system platform. It will be demonstrate that it is easy-to-operate, suitable to perform TPFE, SHG, and THG imaging, using only the standard femtosecond Ti:sapphire laser. Also, the modifications over the scan head and the detector will be explained and besides the physical evidences of the higher harmonics will be submitted. Following, the usefulness of these techniques to detect early changes associated with the progression of epithelial cancer shall be showed. Thereby, this multi-contrast nonlinear optics microscopy setup, emerge as a valuable tool to differentiate between cancerous, benign, and healthy tissue.

Chapter 4 - Optical microfibers (MFs) are fabricated by adiabatically stretching conventional optical fiber thus preserving the original optical fiber dimensions at their input/output pigtails, allowing ready splicing to standard fibers. MFs diameters, close to the wavelength of propagating light, are of interest for a range of emerging fiber optic applications since they offer a number of enabling optical and mechanical properties including large evanescent fields, high nonlinearity, extreme flexibility and configurability, low-loss interconnection to other optical fibers and fiberized components. However, its application will still be limited because MFs are generally circular symmetric. Here the authors review one kind of novel MF with non-circular symmetric geometry, the fiberized slot waveguides (FSW) or slot-microfiber (slot-MF), which is a highly birefringent engineered MF with nano-scale slots where light field can be confined and enhanced. The slot-MF has the advantages of both a MF and a slot waveguide. Taking advantage of the unique and flexible geometry, a number of significant linear and nonlinear characteristics can be realized including high birefringence, ultra-flattened and low dispersion profiles, and strong surface second-order nonlinearity. With the incorporation of high-index fiber and high nonlinearity material inside the slot, slot-MFs should be ideally suited for a host of nonlinear optical interactions and devices, and offers a large potential in miniature fiber devices for all-optical, fast operating functions for optical telecommunications.

Chapter 5 - Over the past few years significant progress has been made in investigations into the synthesis of nanowires of various materials and their applications to nanoscale electronics, optical devices and sensing systems; nevertheless, a major hurdle facing the developers of the integrated nanowire systems is organizing, manipulating, or controlling the placement of large numbers of nanowires. These problems can be solved by using a shadow deposition method for fabrication of polycrystalline metallic nanowires. This technique uses topographical features of ridge-to-valley to expose specific areas of the alkali halide substrates in order to grow the nanowire arrays. Nanowires of large areas show a preference for formation along exposed steps in the atomic structure of the substrate. The minimum widths of the obtained nanowires are less than 40 nm.

The strongly geometrical shape, high aspect ratios, and nanoscale cross-section of metallic nanowires can be expected to affect optical properties through confinement effects. Optical properties of the obtained metallic nanowires are then investigated by studies of linear reflectivity spectra, linear absorption spectra, and second harmonic generation spectra. Theoretical calculation by Maxwell-Garnett model and finite-difference time-domain method are also represented for predicting the reflectivity, absorption and the electric field distributions close to the wires. With experimental and theoretical studies, it is discovered that there are differences of optical properties among these metallic nanowires. These will be one of the most important issues when considering the types of materials used in current applications and development of new applications.

Chapter 6 - Nowadays it is interesting the huge amount of ionic liquids research that involves their green chemistry and their physical and chemical properties, whilst there has not been a deep research involving their potential applications in optical properties and optoelectronic devices. In this chapter, the authors show a review of recent research results on this topic, from nonlinear optical properties to potential applications; theoretical and experimental results of common ionic liquids. The authors also present new experimental results in NLO properties of six ionic liquids. Finally, the authors summarize some ionic liquids applications in the optical and optoelectronic areas.

In: Nonlinear Optics
Editor: Fred Wilkins

ISBN: 978-1-63321-926-7

Chapter 1

IMPACT OF BANDGAP ON INFRARED OPTICAL NONLINEARITY IN NOVEL QUATERNARY CHALCOGENIDES: Cu_2CdSnS_4, α/β-Cu_2ZnSiS_4 AND Li_2CdGeS_4

J. A. Aitken,[1,*] J. A. Brant,[1] D. J. Clark,[2] Y. S. Kim[2,3] and J. I. Jang[2,+]

[1]Department of Chemistry and Biochemistry,
Duquesne University, Pittsburgh, Pennsylvania, US
[2]Department of Physics, Applied Physics and Astronomy,
State University of New York (SUNY) at Binghamton, New York, US
[3]Department of Physics and Energy Harvest-Storage Research Center,
University of Ulsan, Ulsan, South Korea

ABSTRACT

Diamond-like semiconductors (DLSs), which are inherently noncentrosymmetric, exhibit a wide range of compositional flexibility and thus comprise an attractive platform for physical property tuning for an array of potential nonlinear optical (NLO) applications. Although ternary DLSs such as $AgGaS_2$, $AgGaSe_2$ and $ZnGeP_2$ have long dominated the list of benchmark NLO materials, only recently have investigations into the NLO properties been initiated for quaternary DLSs, which offer a higher degree of versatility. We have studied the

[*] Corresponding authors: E-mail address: aitkenj@duq.edu & jjoon@binghamton.edu.

efficiencies of second harmonic generation (SHG) and third harmonic generation (THG) as a function of broadband wavelength for several relatively unexplored quaternary DLSs. Cu_2CdSnS_4 adopts the stannite structure that is based on the cubic diamond structure, crystallizing in space group I-*42m*. α- and β-Cu_2ZnSiS_4 exist as two diamond-like polymorphs crystallizing in the wurtz-stannite (P$mn2_1$) and wurtz-kesterite (P*n*) structures, respectively, which are derived from the hexagonal lonsdaleite structure. Li_2CdGeS_4 forms with the wurtz-stannite structure. These materials exhibit relatively high thermal stability and are air stable. The selection of elements used to construct these materials plays a key role in the resulting physicochemical properties such as bandgap energy. Electronic structure calculations provide insight into how the composition affects the magnitude of the bandgap, which in turn influences the optical transparency window and laser-damage threshold (LDT). The energy gaps of these semiconductors range from ~ 0.9 to 3.2 eV. Typically, enhanced second-order nonlinearity has been expected for relatively narrow-gap materials, but highly efficient generation of long wavelengths requires a sufficiently wide bandgap. The narrow-gap semiconductor Cu_2CdSnS_4 exhibits high SHG and THG responses in the IR, but practical applications are limited by LDT, significant one-photon absorption at 1064 nm. On the other hand, the relatively wide-gap semiconductors α- and β-Cu_2ZnSiS_4 exhibit a high LDT and a wide transparency window, but relatively low SHG and THG responses. Li_2CdGeS_4, an especially unique material, possesses a wide bandgap and exhibits high SHG and THG efficiencies along with an exceptional LDT, demonstrating that high coefficients of nonlinearity and wide bandgaps are not mutually exclusive.

1. INTRODUCTION

1.1. Principles of Nonlinear Optics: Harmonic Generation, Optical Kerr Effect and Multiphoton Absorption

Every nonlinear optical (NLO) phenomenon occurs as a consequence of a change in the optical properties of a host material in the presence of a strong electromagnetic perturbation, which in turn induces the modification of the light field, namely light-matter interaction. The corresponding NLO efficiency therefore critically depends on the material properties that also determine nonlinear dispersion via wavelength dependence.

When intense light enters an NLO medium, dielectric polarization of the medium responds *nonlinearly* to the electric field, $E(t)$, of the incident light at time t and can be described by

$$P(t) = \varepsilon_0[\chi^{(1)}E(t) + \chi^{(2)}E^2(t) + \chi^{(3)}E^3(t) + \cdots], \quad (1)$$

where ε_0 is the vacuum dielectric constant and the coefficients $\chi^{(n)}$ are the n-th order susceptibilities of the NLO material. Although Eq. (1) is written in its simplest form, $\chi^{(n)}$ are generally tensors of rank $R = n + 1$ operating on each of the components of the electric field.

Second harmonic generation (SHG) arises from the $\chi^{(2)}$ term in Eq. (1) and is nonzero only in crystals which lack inversion symmetry (*noncentrosymmetric*). SHG results in the phenomenon of frequency doubling, where an input pump wave generates a wave with twice the frequency, and therefore half the wavelength, inside an NLO material. Under negligible pump depletion and in the slowly varying envelope approximation, the expression for the SHG intensity, $I(2\omega, d)$ with d being the interaction length, is given by

$$I(2\omega, d) = \frac{\omega^2|\chi^{(2)}|^2 d^2}{2n_{2\omega}n_{\omega}^2 c^3\varepsilon_0}\operatorname{sinc}^2\left(\frac{\Delta kd}{2}\right)I^2(\omega) \propto |\chi^{(2)}|^2 I^2(\omega), \quad (2)$$

where n_ω and $n_{2\omega}$ are refractive indices at ω and 2ω, respectively, c is the speed of light in a vacuum and $I(\omega)$ is the pump intensity. Note that Eq. (2) is maximized for the so-called *phase-matching* condition, $\Delta k = k(2\omega) - 2k(\omega) = 0$, and the corresponding SHG intensity increases quadratically with d, i.e., the process is "accumulative" [1]. In contrast, if the process is not phase-matchable, SHG oscillates as $\operatorname{sinc}^2(\Delta kd/2)$, being out of phase with the incident field for $d > l_c = \pi/\Delta k$ (coherence length). Thus, the SHG intensity cannot exceed $I(2\omega, l_c)$ and there is no benefit to the use of an NLO material of a thickness larger than l_c. Potential second-order NLO materials therefore must support phase-matching ($l_c \to \infty$).

Efficient SHG conversion can be achieved by employing various techniques, such as intracavity resonant conversion [2,3], nonlinear waveguides [4,5] and periodic poling [6-8]. In fact, wave mixing $(\omega_1 \pm \omega_2)$ of two different frequencies, ω_1 and ω_2, has been used to generate a wide range of wavelengths using crystals with suitable $\chi^{(2)}$ values. *Difference frequency generation* (*DFG*) is especially important for generating mid-infrared (IR) and terahertz (THz) radiation [9,10], and facilitates multichannel conversion and all-optical networking [11].

Similarly, third harmonic generation (THG) is the frequency tripling process that can be achieved with third-order nonlinearity, the $\chi^{(3)}$ term in Eq. (1). Although THG-based microscopy, spectroscopy and frequency-resolved optical gating have been proposed [12,13], the most important third-order NLO applications involve the optical Kerr effect and multiphoton absorption (MPA) as explained below. In general, third-order nonlinearity is characterized by Z-scan [14, 15], but the technique cannot be employed for pristine powdered samples. Therefore, a powder THG method [16] was employed to assess $\chi^{(3)}$ of our samples in this chapter. Under the same conditions, the corresponding THG intensity, $I(3\omega, d)$, is

$$I(3\omega, d) = \frac{9\omega^2 \left|\chi^{(3)}\right|^2 d^2}{16 n_{3\omega} n_{\omega}^3 c^4 \varepsilon_0^2} \operatorname{sinc}^2\left(\frac{\Delta k' d}{2}\right) I^3(\omega) \propto \left|\chi^{(3)}\right|^2 I^3(\omega), \tag{3}$$

where $n_{3\omega}$ is the index of refraction at 3ω and $\Delta k' = k(3\omega) - 3k(\omega)$ is the THG phase mismatch. Generally, materials showing high $\chi^{(2)}$ responses also exhibit high $\chi^{(3)}$ responses [1,16]. Unlike SHG, however, THG is possible in *centrosymmetric* and even glassy materials [16-18]. THG is typically non-phase-matchable due to a large $\Delta k'$, but third-order NLO devices do not require beam mixing; thus, phase-matching is not particularly required.

Harmonic generation and relevant nonlinearity can be characterized by the Maker fringe method [19] or Kurtz powder method [20], where the former is only applicable to samples prepared in thin-film or bulk crystalline forms. These techniques are mostly available at fixed wavelengths. However, nonlinearity strongly depends on the input wavelength, $\lambda = 2\pi c/\omega$. This implies that broadband NLO characterization is required for the accurate assessment of potential NLO materials. Therefore, our broadband NLO capability, as explained in Sec. 3, is ideal for investigating important NLO processes over an exceptionally wide wavelength range; thus, it can characterize fundamental NLO dispersions, as well as practical working ranges of the NLO materials.

Many state-of-the-art NLO devices utilize third-order NLO properties involving optical switching and amplification [21,22]. Namely, the optical Kerr effect and 2-photon absorption (2PA) result from the real and imaginary parts of $\chi^{(3)}$, respectively, which are related by the Kramers-Kronig relations [1]. The optical Kerr effect arises from the modification in the refractive index,

n, of an NLO medium in the presence of intense light: $n = n_0 + n_2 I(\omega)$, where n_0 is the linear refractive index, $n_2 = 3\chi^{(3)}/4n_0^2\varepsilon_0 c$ is the NLO index, and $I(\omega)$ is the intensity of the incident light. The corresponding refractive index change is proportional to $I(\omega)$ and, therefore, the absolute square of the local field of light. The value of n_2 depends on the material itself and can range from 10^{-6} cm^2/W to 10^{-16} cm^2/W depending on the origin of nonlinearity [1]. Here we are interested in the *bound electronic response* where our THG approach can assess n_2 of the pristine powdered materials for potential $\chi^{(3)}$ applications. Relevant phenomena involving the Kerr effect include self-focusing, self-phase modulation, temporal solitons, supercontinuum generation and optical bistability [23-27].

MPA arises from simultaneous absorption of photons that induces electronic excitation of a material, where the number of photons, N, involved is determined by $N\hbar\omega \geq E_g$ (bandgap). It is a non-parametric NLO process that depends on the N-th power of $I(\omega)$, and is therefore generally much weaker than linear absorption. Current interest in MPA includes multiphoton spectroscopy and microscopy [28-30], which has the advantage of enhanced spatial resolution. Although MPA is also important for optical limiting applications [31-33], MPA-induced *nonlinear loss* and subsequent *optical damage* can significantly limit the efficiency of various NLO devices. In this chapter, we present both second- and third-order NLO properties of diamond-like chalcogenide semiconductors, Cu_2CdSnS_4, α/β-Cu_2ZnSiS_4 and Li_2CdGeS_4, as a function of wavelength and study the effects of bandgap on optical nonlinearity as well as laser-damage threshold (LDT).

1.2. Applications of IR NLO Materials

NLO materials provide a means for wavelength conversion, optical amplification, optical phase conjugation and switching in the ultraviolet (UV), visible, IR and THz regions. The compounds studied in this chapter were particularly investigated for the generation of radiation in the IR regime, as many applications require improved materials at these wavelengths. Generally speaking, IR NLO materials benefit military, medical, commercial and industrial applications. The development of all-solid-state lasers that use these NLO crystals is especially important for military and medical applications because they are relatively light and compact, allowing ease of transportation and use [34,35].

The military drives NLO research and development of improved materials to further expand upon information technology systems, laser weapons and laser radar. Advancements in information technology, which benefit not only the military but also the commercial and industrial sectors, can aid the military in data storage, displays and communication, such as laser communication for submerged submarines [36]. The military has a need for dynamic IR scene projectors for simulation training, as well as for the evaluation of systems and components, such as missile testing and the simulation of very hot objects [37-39].

While current, mature NLO materials are applicable to low-power (P_{avg} <1 kW) weaponry, the lack of necessary power handling ability and efficiency inhibits the use of these materials for high-power (P_{avg} > 1 kW) applications used in aircraft-, ship- or tank-based systems [34]. IR-laser sources are useful in IR countermeasure (IRCM) systems employed to defend military vehicles, particularly aircrafts, by essentially "jamming" the missile's IR guidance system [34,40]. Only NLO materials can generate the wavelengths and power necessary for these IRCM systems to operate. For example, the common IRCM system developed by BAE systems is based on a $ZnGeP_2$ optical parametric oscillator (OPO). These systems have even been considered for use on commercial airliners [35]. Yet there is a need to develop these systems further for improved durability, power and efficiency.

Laser radar can be used for target acquisition, tracking and pointing. Future systems are desirable to operate at eye-safe wavelengths beyond 1.5 µm, an area where the coverage of current solid-state lasers is restricted [41,42]. Furthermore, using vibrational spectroscopy for the remote sensing of chemical and biological weapons [43], as well as stand-off detection of drugs and explosives at the parts per trillion level, requires access to the mid-IR, also known as the molecular fingerprint region. Trace detection of explosives, such as 2,4,6-trinitrotoluene (TNT) and cyclotrimethylenetrinitramine (RDX), has been demonstrated with mid-IR cavity ring-down spectroscopy (CRDS) using a broadly tunable (6–8 µm) $ZnGeP_2$ OPO [44]. The wavelengths used in these devices must be chosen with great precision so that the detection system is highly selective for a particular species; however, sensitivity and specificity are areas that need further improvement in these prototypes. Furthermore, higher power sources and better tuning ranges in the mid-IR are needed [34].

In the medical field, IR NLO materials have applications in diagnostics, monitoring, treatment and surgery. Noninvasive diagnosis and early detection of certain diseases and metabolic disorders can be conducted via breath analysis by molecular spectroscopy in the region of 2–20 µm. Mid-IR CRDS using an SHG material is a feasible method for the diagnosis of cystic fibrosis (CF), which currently requires invasive procedures. Most adults suffering from CF are infected by a specific bacterium that produces hydrogen cyanide, which can be detected by the spectrometer [45]. The detection of other molecules can be indicative of various medical problems, for example alkanes in the case of lung cancer and formaldehyde in the case of breast cancer [45,46]. IR NLO materials provide a means for noninvasive medical monitoring. For example, the blood sugar level in a diabetic patient can be continually assessed by passing IR light through a person's ear lobe and measuring the attenuation at the glucose resonances near 9 μm [47-49].

A number of devices can be envisioned using IR NLO materials through DFG, SHG or optical parametric oscillation. For example, solid-state molecular sensors (SSMS) are one type of chemical and biological molecule detection system. A wavelength tuning element, a key component of the SSMS device, is built up of a birefringent chalcopyrite layer surrounded by optically confining and guiding cladding layers, for example, optically confined $Zn(Ge_{1-x}Si_x)P_2$ layers on GaP and silicon substrates [50]. Among the criteria for these detection systems are the ability to operate at normal environmental conditions, functional reliability, long lifetime and portability [50]. Further advancements of molecular detection systems rely on stringent control of composition and point defect chemistry in the NLO chalcopyrite constituent [50].

Medical laser treatment and surgery in the areas of dermatology, ophthalmology and general medicine also make use of IR NLO materials [51]. For example, certain malignant human tissue absorbs radiation at ~8 μm, while the adjoining normal tissue absorbs much less [52]. The generation of tunable mid-IR frequencies in the range of 5.5–9.3 μm could be useful for this application and has been demonstrated using a 2-μm pumped $ZnGeP_2$ OPO [52]. A $ZnGeP_2$ OPO operating at 6.45 μm is promising for use in high-precision laser surgery since it imparts minimal adverse thermal effects by exploiting the amide-II vibration wavelength. With this design, most of the collateral damage to the surrounding tissue is avoided since proteins, but not water, are heated [51].

There are also numerous applications of IR NLO materials in the commercial and industrial sectors. Atmospheric monitoring of air [53,54] and water pollutants can be conducted based on the vibrational signatures of the molecular species in the IR region; for instance, the strongest vibration of methane is at 3.39 μm and carbon monoxide at 4.67 μm [47]. A tunable diode-laser spectrometer for the mid-IR region (~7.2 μm) has been constructed using

DFG with $AgGaS_2$ [55]. This region of the mid-IR is especially useful for trace-gas detection, in particular sulfur dioxide, ozone and nitric oxide, among others [55]. The oil and gas industry can benefit from leak detection using similar technology. Other areas of applications include industrial process control [56], materials processing [57], optical frequency metrology [58], wind profile analysis, and climate study [59]. NLO materials offer increased speed and contribute to lighter and more compact devices in the information technology realm. More efficient NLO materials with higher damage resistance are needed for optical communication, data storage and computing using optical integrated circuits [11].

While the signatures of molecules, as per vibrational wavelengths, are found in the IR region, rotational transitions of molecules can be detected in the THz region. Some progress has been made toward a widely tunable THz source for this application using $ZnGeP_2$ in a DFG scheme to generate short THz pulses [60]. THz radiation is also useful in biomedical imaging and security screening. Employing THz radiation in the detection of concealed weapons is viable since many optically opaque materials, such as fabrics and paper, are transparent to THz radiation. Currently, X-ray radiation is used for such purposes but THz radiation is advantageous since it is non-ionizing [61].

1.3. Criteria for Ideal NLO Materials

A number of criteria must be considered in the search for new NLO materials [62]. One obvious quality that is sought after in an NLO material is significant optical nonlinearity. As mentioned in Sec. 1.1, a prerequisite for second-order nonlinearity is that the structure of the compound must be noncentrosymmetric. Third-order nonlinearity does not have this requirement, as the material can be centrosymmetric or noncentrosymmetric. Adequate birefringence due to optical anisotropy is needed in order to achieve phase-matching for efficient wavelength conversion. Access to large crack-free single crystals is important, so that the materials may be sizeable for use in practical applications. To access a wide window of optical transparency, the crystals must be prepared in a manner to minimize absorption losses, which are usually created by defects. Since the vibrations of bonds involving relatively heavy elements occur further into the IR, the use of relatively heavy elements in the compound ensures a wider IR-transparent window. High thermal conductivity, as well as the ability to hold up to strong lasers without significant damage, i.e., high LDT, is crucial for high-powered applications.

LDT usually scales with bandgap, such that wide-bandgap materials have higher LDTs. Other considerations include the phase diagram, environmental stability, mechanical processability and toxicity, among others. Several of these attributes define the wavelengths at which the NLO material can be used. As explained in Sec. 1.2, access to specific wavelengths is necessary for certain applications.

While high NLO susceptibility is important and often reported by researchers who are making new materials, it has been pointed out by Peter G. Schunemann of BAE systems that elevated nonlinear coefficients are less critical than other practical considerations such as small absorption loss, high LDT, low thermal lensing and non-critical phase-matching [63].

1.4. Current benchmark IR NLO materials

Considering the long list of criteria for useful NLO materials, it is not surprising that a single compound that sufficiently meets all of these requirements has not yet been discovered. Furthermore, there is not one material that can function across the entire IR spectrum. Each of the crystals that are currently available, either from commercial sources or from laboratories, has a unique set of strengths and weaknesses. These mature IR NLO materials can be divided into three groups, (1) the diamond-like chalcopyrite (I-*42d*) semiconductors consisting of $CdGeAs_2$, $ZnGeP_2$, $AgGaSe_2$ and $AgGaS_2$, (2) the diamond-like β-$NaFeO_2$ (P*na2₁*) semiconductors such as $LiInS_2$ and $LiInSe_2$, and (3) the layered GaSe ($P6_3/mmc$).

$CdGeAs_2$ is widely known for the enormous $\chi^{(2)}$ value of 472 pm/V [63]. Additionally, it has a wide transparency window, wider than $ZnGeP_2$, which is discussed below. It can be used for efficient DFG of 7 μm to 20 μm [64], but the crystal growth of $CdGeAs_2$ is severely hindered by large anisotropic thermal expansion that often results in cracking [65]. It also suffers from MPA and can only be pumped with low photon energies [66]. Furthermore, the short wavelength cutoff, ~2.3 μm, of $CdGeAs_2$ makes it incompatible with most practical solid-state pump lasers [63], thus the material is not commercially available.

$ZnGeP_2$ is most often used for frequency doubling of a CO_2 laser (10.6 μm) [67]. The compound is commercially available in IRCM systems [34,40] and has demonstrated potential in IR spectroscopy [43,44] and laser surgery [51], as explained in Sec. 1.2. Its appeal lies in the relatively large $\chi^{(2)}$ (150 pm/V) and thermal conductivity of 36 W/mK, the largest for a material that is transparent and phase-matchable at 2 μm [62]. Single crystals of $ZnGeP_2$ have been difficult to grow due to native defects that result in absorption losses; however, improvements have been made through advances in the growth of sizeable crystals using the horizontal-gradient-freeze (HGF) technique [68]. The vertical Bridgman growth technique followed by post-growth annealing in red phosphorus and ZnP_2 powders has also been successful for the preparation of high-quality crystals [69]. Here one must be very cautious in ramping up the temperature slowly, allowing for the mixture to react at lower temperatures, since the high vapor pressure of phosphorous can cause the growth ampoules to explode at high temperatures [62,69]. On the other hand, $ZnGeP_2$ is limited in its applications for CO_2 doubling or long-wave OPOs due to the onset of MPA around 8.5 μm [70]. While improvements have been made, the LDT still precludes the use of $ZnGeP_2$ for extremely high-powered applications where higher efficiencies and output powers are necessary [68].

$AgGaSe_2$ has a major advantage over $ZnGeP_2$ in that it is optically transparent from the visible region all the way out to ~19 μm and the crystals have little absorption losses [71,72]. The phase-matching capability of $AgGaSe_2$ for a broad range of wavelengths is advantageous for a gamut of three-wave-mixing applications [72]. However, these advantages have trade-offs. Compared to $ZnGeP_2$, $AgGaSe_2$ has a weaker nonlinearity ($\chi^{(2)}$ = 66 pm/V) [62] and a thermal conductivity that is 30 times less than that of $ZnGeP_2$ [73]. In addition, it suffers severe anisotropic thermal expansion such that, when the crystals cool, they expand along the *c*-axis but contract along the *a*- and *b*-axes, usually resulting in cracks unless an oriented seed crystal is appropriately implemented [74]. Moreover, like $ZnGeP_2$, MPA is a problem; $AgGaSe_2$ undergoes 2PA at 1 μm that contributes to the less than ideal LDT [75]. The 2PA effect severely limits applications of $AgGaSe_2$ for use in mid-IR DFG [76]. Another issue with $AgGaSe_2$ is that it is non-phase-matchable at 1 μm [63].

$AgGaS_2$ is phase-matchable and transparent at 1 μm [63]. However, this comes at a price in terms of the lower $\chi^{(2)}$ of 36 pm/V [77]. It also has a relatively low LDT that is comparable to that of $AgGaSe_2$ [75]. $AgGaS_2$ is more problematic for extended laser illumination due to so-called photo-darkening [78]. Like $AgGaSe_2$, $AgGaS_2$ possesses low thermal conductivity [73] and undergoes severe anisotropic thermal expansion that makes the growth of large, crack-free single crystals challenging [79].

In an attempt to escape the problems of $AgGaSe_2$ and $AgGaS_2$, such as negative thermal expansion along the optic axis, low thermal conductivity and less than desirable LDT, researchers have recently begun to examine members of a different family of NLO materials with the diamond-like β-$NaFeO_2$ structure, e.g., $LiInS_2$, $LiInSe_2$, $LiGaS_2$ and $LiGaSe_2$ [80,81]. The higher LDTs of these materials, as compared to the chalcopyrites, make them viable contenders for high-power NLO applications. Since $LiInS_2$ and $LiInSe_2$ are the most mature, only those two members are discussed in this chapter.

$LiInS_2$ is one of the widest-gap semiconductors among IR NLO materials [82]. The wide bandgap (> 2.4 eV) serves to essentially eliminate 2PA effects when pumping with the Nd:YAG line (1.064 μm) [83]. Since the structure is not chalcopyrite-like, $LiInS_2$ does not suffer the same degree of anisotropic thermal expansion as $AgGaSe_2$ and $AgGaS_2$. While the thermal expansion coefficients are still slightly anisotropic, they are all positive (along *a*, *b* and *c* axes) [84]. The large thermal conductivity, which is twice that of $LiNbO_3$ and five times larger than that of $AgGaS_2$, is also an advantage [85]. Moreover, the material has a relatively large birefringence and a wide transparency window, and therefore, can be useful at THz frequencies for pulse splitters and pulse shapers [86]. It allows for direct down-conversion of a Ti-sapphire laser (0.8 μm) to the 5–11 μm range in a single step [87]. Yet this material is not flawless. The $\chi^{(2)}$ value, which has been reported as ~7–15 pm/V [85,88-90], is smaller than those of the chalopyrites mentioned above. While anisotropic thermal expansion is not an obstacle in crystal growth, the material is plagued by defects, such as vacancies and anti-sites, resulting in different colored samples [91]. However, some of these problems can be somewhat corrected by post-growth annealing in S vapor; thus, $LiInS_2$ crystals are currently available for purchase. Due in part to the large bandgap, the LDT of $LiInS_2$, determined using the 9.55 μm line and a pulse width of 36 ± 2 ns, has been reported as 180 MW/cm^2, which is significantly larger than those of $AgGaSe_2$ and $AgGaS_2$ under identical conditions [92]. On the other hand, using the Nd:YAG line, it was observed that $LiInS_2$ had an LDT of 100 MW/cm^2 at 1.064 μm (10 ns), which is closer to those of $AgGaSe_2$ and $AgGaS_2$ [82]. Comparable results were also obtained at 5 μm [85]. However, it is anticipated that continued improvements in the crystal growth of this material, resulting in less defects and absorption losses, should improve the LDT [82]. Interestingly, $LiGaS_2$ has an even wider bandgap than $LiInS_2$ and recent work has shown that $LiGaS_2$ possesses a superior LDT compared to $LiInS_2$. However, the $\chi^{(2)}$ value is only 11 pm/V [93] and sizeable crystals need to be developed [94].

The selenium analogue of $LiInS_2$, $LiInSe_2$, offers a collection of attributes that are similar in quality to $LiInS_2$. In comparison to $AgGaSe_2$ and $AgGaS_2$, $LiInSe_2$ has a higher thermal conductivity, a lesser degree of anisotropic thermal expansion and a wider bandgap [95]. The $\chi^{(2)}$ value has been reported as 17 pm/V and 22 pm/V [90,96]. Due to the wide optical transparency window and bandgap, an Nd-YAG-pumped $LiInSe_2$ OPO is capable of operating from 4.7–8.7 μm without 2PA [97]. Further, $LiInSe_2$ has been noted for its superior thermo-mechanical properties [97]. However, like $LiInS_2$, the crystals of $LiInSe_2$ are plagued with discolorations due to point defects that occur during growth, resulting in crystals that are yellow, greenish, rose and dark red [98]. Using the 9.55 μm line with a 30 ± 2 ns pulse width, the LDT of $LiInSe_2$ has been determined to be 248 MW/cm^2, which is noted as being 1.6 to 1.8 times higher than that of $CdGeAs_2$ (157 ± 13 MW/cm^2), $ZnGeP_2$ (142 ± 9 MW/cm^2), $AgGaSe_2$ (139 ± 6 MW/cm^2) and $AgGaS_2$ (149 ± 6 MW/cm^2) [99]. One of the main challenges remaining for $LiInSe_2$ is the reduction of residual losses in the transparency range [100]. Interestingly, $LiInSe_2$ is also a promising candidate for neutron detection applications [101].

GaSe has extremely wide regions of optical transparency and phase-matchability, as well as high second-order nonlinearity [102]. However, the main drawback of GaSe lies in the difficulty to mechanically process the crystals, since the soft, layered material can be cleaved only along the (001) plane [103]. Furthermore, the LDT is not stellar [66]. Attempts to improve the properties of GaSe by doping have met some success; yet, further improvements are required [104].

2. Strategies for the Discovery of New Materials for IR Wavelength Conversion

The discovery of new materials is vital for the advancement of nonlinear optics, especially for applications that require high power and longer wavelengths, throughout the IR and THz frequencies. Many of the approaches used to target improved NLO materials for the IR, specifically *noncentrosymmetric* structures, involve the assembly of acentric building blocks. Although the assembly of specific building blocks is often targeted, other acentric building blocks and noncentrosymmetric structures result from serendipity and exploratory synthesis. Some of the most widely reported building blocks used in the construction of noncentrosymmetric structures are provided along with a few important example materials; but, this is not intended to serve as a complete survey.

Highly asymmetrical octahedral building blocks can be accessed by combining a d^0 transition metal cation, Ti^{4+}, V^{5+}, Nb^{5+}, etc., with a cation that contains a lone pair of electrons, Bi^{3+}, Sn^{2+}, Se^{4+}, etc., for the assembly of noncentrosymmetric oxides [105-107]. The formation of the lone pair of electrons is mediated by the interactions between the *s* orbital of the metal and the *p* orbitals of the oxide. Out-of-center distortions occur as the oxide ligands are effectively "pushed" to one side of the cation by the lone pair of electrons. Second-order Jahn-Teller (SOJT) distortion in octahedrally coordinated d^0 transition metals is considered to be the primary distortive cause [106] in such materials. SOJT distortion arises to remove a host of nearly degenerate electronic configurations within the extended structures that result from combining the empty *d* orbitals of the metal with the filled *p* orbitals of the ligands. The direction and magnitude of the distortion generally depend on the cations employed. For example, Ti^{4+}, V^{5+} and Nb^{5+} usually have out-of-center distortions toward an edge or corner of the octahedron, while Mo^{6+} and W^{6+} are distorted toward an edge or face. The magnitude of distortion generally increases as $Ti^{4+} < Ta^{5+} < Nb^{5+} < W^{6+} < V^{5+} < Mo^{6+}$. Indeed, this approach has resulted in many noncentrosymmetric extended structures, such as $SbNbO_4$ (P*na2₁*) [107], $InNb(TeO_4)_2$ (P*ca2₁*) [108] and even some containing organic ligands (e.g., $HdpaNbOF_4$, dpa = 2,2′-dipyridylamine (C*c*)) [109]. While many interesting materials have been discovered with this approach, most of these oxides are limited for NLO applications in the IR due to absorption issues. Although the rationale for microscopic polarity is established, the driving force for the assembly of the macroscopically acentric extended structure is often uncertain [110].

Another approach employs building blocks that contain stereochemically active lone pairs (SCALP), such as $[SeO_3]^{2-}$ and $[IO_3]^-$ [110-113], for the assembly of noncentrosymmetric structures that have a periodic arrangement of polarizable free electrons. Some of these materials have transparency windows ranging from the visible to the IR (12.5 μm) [114-116], which is a rarity in oxides. Among these is the polar NaI_3O_8 (P-*4*), which is assembled with the $[I_3O_8]^-$ building block that can be viewed as a condensation of three $[IO_3]^-$ units with three stereochemically active lone pairs [111a]. The formation of the targeted building blocks can be synthetically challenging, and this design strategy is afflicted by centrosymmetric contenders. For example, the lone pairs on Pb^{2+} and Se^{4+} cations deem the PbO-SeO_2 system as attractive for generating polar materials. Accordingly, several compounds have been discovered, including $PbSeO_3$, $PbSe_2O_5$, Pb_3SeO_5 and Pb_5SeO_7 [110]. However, only Pb_3SeO_5 is noncentrosymmetric. While these results offer advancements in the discovery of NLO materials, the compositions and resulting properties of materials in these families are limited.

Borates, based on $[BO_3]^{3-}$ units, are a rich source of NLO materials [117], including β-BaB_2O_4 (BBO) and LiB_3O_5 (LBO) that are commercially available and widely used for SHG. The second-order nonlinearity is attributed to the asymmetric electronic distributions on the distorted planar anions, namely $[B_3O_6]^{3-}$ in BBO and $[B_3O_7]^{5-}$ in LBO. These π-delocalized building blocks result from joining BO_3 and BO_4 units. Recently, research efforts have been directed toward quaternary compositions. $Pb_2B_5O_9I$ exhibits exceptional SHG (λ = 1.064 μm), which is attributed to the cooperation of I^-, the lone pair of electrons on Pb^{2+}, and the BO_3 and BO_4 building units; however, $Pb_2B_5O_9I$ is not suitable for applications in the far-IR since the transparency spans from the near-UV to the mid-IR [118].

Combining multiple types of building blocks within a single material can lead to "constructive" polarizations and high SHG efficiencies [111b,115, 117c,119]. For example, $Cd_4BiO(BO_3)_3$ exhibits exceptional SHG (λ = 1.064 μm), as a result of combining polar displacement of d^{10} Cd^{2+} with the lone-pair Bi^{3+} and the $[BO_3]^{3-}$ building unit [117c]. SOJT-distorted d^0 metal centers can be combined with SCALP building blocks to yield noncentrosymmetric materials, such as $BaTeMo_2O_9$ and $TlTeVO_5$ that yield much higher SHG than SiO_2 [120,121].

The cause for the asymmetric building blocks to align into solid-state noncentrosymmetric structures is often unclear. Accessing and controlling the assembly of these building blocks can be challenging, competing centrosymmetric structures are often difficult to avoid, and compositions can be limited. While all of these approaches have led to insightful discoveries, they are mainly used for oxides. BBO and LBO exhibit efficient NLO behavior in the UV region [122,123], while metal oxides, such as $KTiOPO_4$, exhibit efficient NLO behavior only in the visible region [124]. Such oxides generally exhibit higher LDTs, but optical transparency is limited between the UV and near-IR range. On the other hand, *chalcogenides* are prominent for NLO applications in the long-wavelength region due to wider regions of optical transparency and higher nonlinear susceptibilities in the IR [72,125, 126]. The windows of transparency are generally established by bandgap absorption and M-S (M = metal) stretches at frequencies in the far-IR that can be tuned via composition. While the mature NLO chalcogenides have drawbacks, as outlined in Sec. 1.4, the discovery of new NLO chalcogenides is essential for advancements in IR applications. Indeed, novel chalcogenides with noncentrosymmetric structures exhibit attractive properties for NLO applications [126-130]. However, simultaneously optimizing optical nonlinearity and the threshold for laser-induced damage, in addition to other valuable properties, has not been a trivial feat. Systematic studies of the relationships between composition, structure and properties are vital to developing materials design efforts, but are often futile since most new NLO chalcogenides are discovered via exploratory synthesis [130-133].

Chalcogenide analogues (e.g., $[AsS_3]^{3-}$, $[SbS_3]^{3-}$, $[TeS_3]^{2-}$) of the oxide building blocks (e.g., $[IO_3]^-$, $[TeO_x]^{n-}$, $[BO_3]^{3-}$ and $[B_3O_6]^{3-}$) for NLO materials are relatively unexplored [134-136]. Finding the synthetic conditions to facilitate the formation and assembly of such building blocks into noncentrosymmetric structures is challenging, but there are a few examples in which these types of materials have been accessed. The combination of $[TQ_4]^{n-}$ tetrahedra and acentric $[SbQ_3]^{3-}$ trigonal pyramids, in which T is a group 13 or 14 element (e.g., Al, Ga, Ge) and Q is a chalcogen (i.e., S, Se or Te), has been proposed to assemble new noncentrosymmetric materials [137]. Indeed, SbS_3 pyramids with stereochemically active lone pairs and GaS_4 tetrahedra render the $Ba_{23}Ga_8Sb_2S_{28}$ structure noncentrosymmetric ($Cmc2_1$). Although the polarity contribution of the SbS_3 pyramids is negligible, they serve to disconnect the GaS_4 tetrahedra, which in turn effectively yields an SHG response which is non-phase-matchable at 2.05 μm [138]. Pyramidal $[AsS_3]^{3-}$ units have been observed in $Li_{1-x}Na_xAsS_2$, which reportedly has an SHG efficiency that is ~30× $AgGaSe_2$ at 1580 nm; however, it is non-phase-matchable. While this material melts congruently, the thermal stability is relatively low (m.p. = 480 °C) [129]. In another example, pyramidal AsS_3 units connect asymmetric $[Ta_2S_{11}]^{6-}$ units in $^1/_\infty[Ta_2AsS_{11}{}^{3-}]$ chains that comprise the acentric $K_3Ta_2AsS_{11}$ and $Rb_3Ta_2AsS_{11}$. Both have SHG efficiencies at 1200–1700 nm that are higher than that of the $AgGaSe_2$ reference sample, but also have relatively low thermal stabilities (m.p. = 400–500 °C), melt incongruently and are non-phase-matchable at 1540 nm [134]. The $^1/_\infty[Ta_2AsS_{11}{}^{3-}]$ chains pack into a centrosymmetric structure when Cs^+ is implemented (i.e., $Cs_3Ta_2AsS_{11}$). While these materials exhibit some interesting optical nonlinearities, they are inadequate to compensate for current shortcomings. These building blocks are difficult to access on demand, and acentric structures are limited to a narrow set of compositions.

Another approach is based on solid solutions, such as the formation of $(I\text{-}III)_{1-x}(IV)_xVI_2$ solid solutions of ternary I-III-VI_2 chalcopyrites and binary IV-VI_2 chalcogenides [139]. For example, the noncentrosymmetric $AgGaGeS_4$ (i.e., $Ag_xGa_xGe_{1-x}S_2$ at $x = 0.5$) was discovered by exploring the solid solution of the $AgGaS_2$ and GeS_2 parent compounds with the goal of increasing the birefringence and transparency range for $AgGaS_2$. Much like $AgGaGeS_4$, $AgGaGeSe_4$ is a member of the $AgGaSe_2$-$GeSe_2$ system [140]. Interestingly, adding IV-VI_2 to the I-III-VI_2 chalcopyrites resulted in increased birefringence, widened bandgaps and higher LDT in comparison to the benchmark materials, in addition to wide optical transparency and phase-matchability at 1 μm [76,141]. These materials are attractive for mid-IR applications (e.g., $AgGaGeS_4$ for DFG of 2–12 μm) [142], but would be enhanced for use at even longer wavelengths if larger transparency regions could be accessed (e.g., $AgGaGeS_4$ has a transparency range of ~0.42–12 μm) and higher LDTs could result from wider bandgaps (e.g., $AgGaGeS_4$ has a bandgap energy of 2.80 eV) [142]. In fact, a wide-gap analogue, $Li_2Ga_2GeS_6$, is more resistant to laser damage, but the clarity window is comparable to that of $AgGaGeS_4$ [125]. Such limitations establish the need for a higher degree of property tuning. By definition, a solid solution is comprised of members possessing a crystal structure that is unchanged with some degree of solubility (e.g., $AgGa_xIn_{1-x}Se_2$); thus, solid solutions are generally formed using parent compounds that are isostructural. Instead, I-III-VI_2 and IV-VI_2 materials have different structures. $AgGaGeS_4$ and $AgGaGeSe_4$ have narrow regions of solubility and lie close to steep phase boundaries within the phase diagrams, which greatly limits the tunability of these compounds [63]. In other works, solid solutions of chalcopyrites have been explored. The phase-matching conditions and refractive index have been tuned as a function of x in the $LiIn(S_{1-x}Se_x)_2$ solid solution [143]. In another example, the phase-matching and birefringence of $AgGaSe_2$ was tuned through substitutional doping of the III-ion in $AgGaSe_2$ to form the $AgGa_{1-x}In_xSe_2$ solid solution [127]. However, the controlled growth of optical-quality crystals of the members of these solid solutions has been deemed "unfeasible", since the Ga:In distribution within the single crystals is not homogeneous [63]. While the discovery of these materials has yielded insight, an optimal collection of properties that are essential for NLO applications in the IR will result from systematic research on composition-property relationships for a class of materials with greater versatility.

As a matter of fact, there exists a robust design strategy for noncentrosymmetric materials in which a high degree of compositional flexibility allows properties to be tuned, and it is easily employed for the discovery of new chalcogenides. Namely, diamond-like semiconductors (DLSs) are accessed by employing tetrahedral building blocks that are predisposed to align along a single direction, rendering the structures noncentrosymmetric. In fact, diamond-like structured materials dominate the list of NLO benchmark crystals for use in the IR. Diamond-like structures are comprised of tetrahedral cations and anions; thus, all selected elements should have a tendency to exhibit tetrahedral bonding and also satisfy the sharing coefficient of 4 for tetrahedral structures, which is the number of tetrahedra participating in the sharing of the tetrahedral corners, Figure 1 [144]. The sharing coefficient is dictated by the electrostatic valence principle, which states that the electronic charge of each anion must be compensated by the cations in the first coordination sphere of the anion [145]. This principle implies that for C_mA_n the number of cations, *m,* must be equal to the number of anions, *n*, to construct a diamond-like structure. The average valence electron concentration (VEC) should be equal to 4. Since cations ideally donate valence electrons to the anions, the VEC per anion (VEC_A) should be equal to 8. Using these guidelines, all of the possible formulae of binary, ternary and quaternary diamond-like materials can be calculated. When the design principles are in play, the formation of diamond-like structures is so favorable that the formation of centrosymmetric contenders is largely uncommon. The assembly of the macroscopically acentric diamond-like structures is likely mediated by efficient packing; cubic closest packing gives rise to cubic diamond, and hexagonal closest packing yields hexagonal diamond, also known as lonsdaleite.

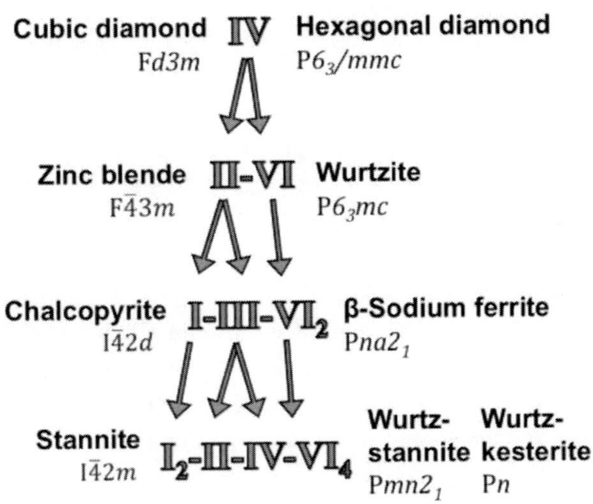

Figure 1. Hierarchy in the structural derivatization of diamond-like structures.

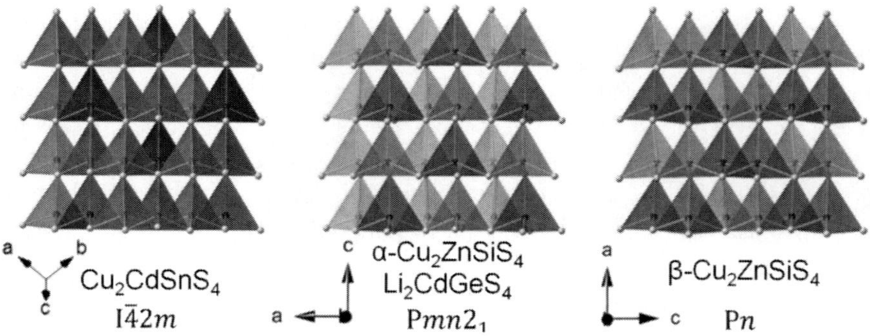

Figure 2. (Color online) I_2-II-IV-VI_4 DLSs with three structure types. I = Cu, Li is green; II = Cd, Zn is blue; IV = Ge, Si, Sn is red; VI = S is yellow.

Although this approach dates back at least to the 1960s [144,146-148] and Pamplin called for "crystal growth and characterization [to] continue in as many laboratories as possible" in 1981 [149], the full potential of DLSs is surprisingly under-investigated. It is even more surprising that Levine's statement from 1973, that the (diamond-like) chalcopyrite structure is so favorable for NLO properties that it will be difficult to find materials with larger nonlinearities in the IR region [150], has been viewed as prophetic [62]. Extending the hierarchy, from ternary DLSs to quaternary DLSs, allows more freedom for property tuning via composition manipulation. Quaternary DLSs are the logical successors of the ternary chalcopyrites that transformed NLO applications for the IR.

Using the guidelines for design, nine quaternary formulae (e.g., II-III_2-IV-V_4 and I-III-IV_2-V_4) have been calculated for normal diamond-like structures and additional formulae are predicted for defect diamond-structured materials with built-in vacancies, (e.g., I-III-IV-□-VI_4 and I_3-IV_2-V-$□_2$-VI_8) [149]. A large number of element combinations can be employed to target diamond-structured materials [144,146]. The I_2-II-IV-VI_4 formula of diamond-like structures is attractive since chalcogenides and alkali metals (e.g., Li) can be simultaneously incorporated while many divalent and tetravalent ions can be selected for property tuning. Toward the goal of designing advanced materials, DLSs are an optimal platform to systematically assess relationships between composition and properties due to the innate structural simplicity and compositional flexibility.

In this chapter, a study of the NLO properties of three quaternary diamond-like systems, namely Cu_2CdSnS_4, α/β-Cu_2ZnSiS_4 and Li_2CdSnS_4, is described, Figure 2. Cu_2CdSnS_4 crystallizes in the noncentrosymmetric space group I-*42m* and has the stannite structure [151-153], which is generated through substitutional derivation of cubic diamond. Cu_2ZnSiS_4 crystallizes in two different diamond-like structures, wurtz-stannite (α-Cu_2ZnSiS_4) and wurtz-kesterite (β-Cu_2ZnSiS_4) in the P*mn2*$_1$ and P*n* space groups, respectively [154,155]. Both polymorphs of Cu_2ZnSiS_4 are structurally derived from lonsdaleite and differ only in cation ordering. Li_2CdGeS_4, like α-Cu_2ZnSiS_4, has the wurtz-stannite structure [156]. The structural changes that result from the compositional variations within these examples of I_2-II-IV-VI_4 DLSs are slight, but the effects on electronic structure can be significant and closely tied to properties that are important for practical NLO applications.

3. Experimental Methods for Nonlinear Optical Measurements

Polycrystalline Cu_2CdSnS_4, Li_2CdGeS_4 and a mixture of α/β-Cu_2ZnSiS_4 were sieved into discrete particle size (*d*) ranges of 2–20 μm, 20–45 μm, 45–63 μm, 63–75 μm, 75–90 μm, and 90–106 μm, in order to assess phase-matching behavior of the samples. In addition, α/β-Cu_2ZnSiS_4 was also sieved into particle size ranges of 106–125 μm and 125–150 μm. Although the samples appear to be air stable, they were placed into fused-silica capillary tubes that were flame-sealed under vacuum to prevent exposure to air and moisture during the measurements. Each tube was loaded into a homemade sample holder that was mounted on a Z-scan translation stage. The absolute nonlinear susceptibilities of the samples were directly compared with those of a reference material, $AgGaSe_2$, prepared by a similar synthetic method [157]. Although α/β-Cu_2ZnSiS_4 is a mixture, the NLO responses measured across various sample spots were reasonably consistent, indicating that the sample is homogeneous. Hereafter, this polymorphic mixture is referred to as Cu_2ZnSiS_4.

Broadband NLO experiments were conducted at room temperature. In order to generate tunable pulses, coherent light of 1064 nm was first produced using an EKSPLA PL-2250 series diode-pumped Nd:YAG laser with a pulse width of $\tau = 30$ ps and a repetition rate of 50 Hz. The Nd:YAG laser pumped an EKSPLA Harmonics Unit (HU) H400, where the input beam was frequency tripled by a successive cascade of NLO wave mixing. The 1064-nm and frequency-tripled radiation (355 nm) from the HU entered an EKSPLA PG403-SH-DFG OPO composed of four main parts; a double-pass parametric generator, a single-pass parametric amplifier, a second harmonic generator, and a DFG scheme. The output wavelengths of the OPO used ranged from 1100 nm to 3300 nm at increments of 200 nm. The corresponding SHG and THG wavelengths are $\lambda_{SHG} = \lambda/2 = 550–1650$ nm and $\lambda_{THG} = \lambda/3 = 367–1100$ nm, respectively.

The incident pulse energy was tuned to 17 ± 2 μJ before being mildly focused onto samples with a spot size of 0.35 ± 0.15 mm in diameter by a CaF_2 convex lens. Here we determined the beam spot size in order to (1) efficiently generate and properly average the harmonic generation signals from powders of random orientations, and to (2) minimize the change in the spot size when λ was varied over a broad experimental range; the beam waist w_0 at the focus undergoes a significant λ-dependent variation via $w_0 = (\lambda/\pi)(f/\sigma)$, where f and σ are the focal length and the Gaussian width of the incident beam [158]. The NLO signals from the samples were collected using reflection geometry and a fiber optic bundle, which was coupled to a selective-grating (1800, 600, and 300 grooves/mm) spectrometer equipped with a charge-coupled device (CCD) camera (Synapse) as well as an extended InGaAs (Symphony) detector. We confirmed that surface-induced effects, as well as NLO signals from other optical components, were negligible. Any thermal load on the samples by the laser pulses tuned below their bandgaps was negligible due to its slow repetition rate of 50 Hz. The relative SHG and THG signals, spectrally resolved in a broad wavelength range, were precisely calibrated with the known and measured efficiencies of all optical components.

In order to estimate LDTs of the compounds, SHG measurements were carried out on the samples of $d = 90–106$ μm as a function of laser intensity in the range of 0.3–17 GW/cm^2 at $\lambda = 1064$ nm and $\tau = 30$ ps. This wavelength is typically used for mid-IR generation via DFG by mixing with a *longer* wavelength, and therefore, the frequency-conversion efficiency can be indirectly estimated by measuring the LDT at 1064 nm. The LDT of a material significantly depends on the pulse width and here it is a picosecond case. LDT measurements based on the powder technique are well established [157,159] and viable principally because the powder size of ~100 μm is much larger than the laser wavelength, i.e., each powder is a bulk itself. This macroscopic nature of powders causes MPA, similar to that found in a large bulk crystal, which is the main mechanism for optical breakdown of NLO materials.

4. SECOND-ORDER OPTICAL NONLINEARITY

The second-order nonlinear susceptibilities of Cu_2CdSnS_4, Cu_2ZnSiS_4 and Li_2CdGeS_4, as well as the $AgGaSe_2$ reference, were determined over a broad fundamental range, Figure 3 [153,157]. It should be noted that all data exhibit a "dip" near λ_{SHG} = 950–1050 nm that is attributed to linear absorption of the fundamental beam by the quartz sample holders [160]. Since Cu_2ZnSiS_4 and Li_2CdGeS_4 have relatively wide bandgaps near the border of the UV and visible regions, bandgap absorption effects are minimal within the measured broadband region. However, Cu_2CdSnS_4 and $AgGaSe_2$ have narrower bandgaps and efficiently absorb the fundamental beam at lower wavelengths. At longer wavelengths, in which bandgap absorption is negligible for all of the compounds, the performances of Cu_2CdSnS_4 and Li_2CdGeS_4 are comparable to $AgGaSe_2$. Such broadband measurements are important for assessing the prospects of new NLO materials since the SHG counts can vary drastically as a function of λ, which is especially evident in Cu_2CdSnS_4.

Phase-matching behavior was analyzed by measuring the SHG responses of the DLSs that were sieved into discrete particle sizes. The phase-matchability of Cu_2CdSnS_4 and Cu_2ZnSiS_4 initiates in the near-IR and phase-matching of Li_2CdGeS_4 starts at the border of the visible and near-IR. The phase-matchability of all samples likely extends into the far-IR since the SHG responses are fairly constant at longer λ; thus, they have wider regions of phase-matchability than $AgGaSe_2$ and are competitive with other commercially available NLO materials for the IR. The phase-matching range of Cu_2CdSnS_4 ($\geq$ 2.1 μm) is wider than the benchmark $AgGaSe_2$, while Cu_2ZnSiS_4 and Li_2CdGeS_4 ($\geq$ 1.7 μm and $\geq$ 1.5 μm, respectively) have wider regions of phase-matchability than $AgGaSe_2$ and $AgGaS_2$, Table 1. Most studies of new chalcogenides with potential IR applications report SHG and phase-matchability studies conducted at a single λ, or for a small range of λ. The wide regions of phase-matchability observed for Cu_2CdSnS_4, Cu_2ZnSiS_4 and Li_2CdGeS_4 are promising for wave-mixing applications [160,162].

To determine $\chi^{(2)}$, the SHG responses were compared to the reference ($AgGaSe_2$) at static regions for which both the reference and the title compounds exhibit minimal absorption and are phase-matchable, Table 1. While Cu_2ZnSiS_4 exhibits the weakest SHG response, lower than that of $AgGaS_2$, the second-order susceptibilities of Cu_2CdSnS_4 and Li_2CdGeS_4 are both higher than $AgGaS_2$ and comparable to $AgGaSe_2$, Table 1. It is not surprising that the narrow-gap semiconductor Cu_2CdSnS_4 yields the highest second-order nonlinear susceptibility and the wide-gap Cu_2ZnSiS_4 system exhibits the lowest SHG response [77]. This phenomenon will be discussed in more detail in Sec. 8.

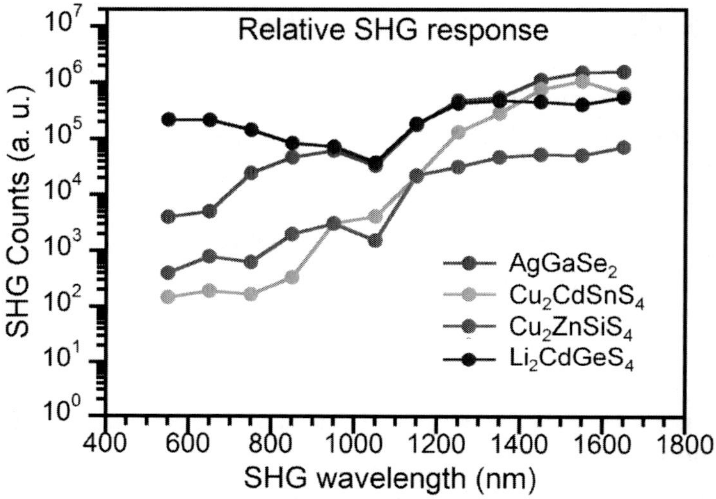

Figure 3. (Color online) Semi-log plot of relative SHG counts from $AgGaSe_2$ (red), Cu_2CdSnS_4 (green), Cu_2ZnSiS_4 (purple) and Li_2CdGeS_4 (blue) with $d = 90$–106 μm.

Table 1. $\chi^{(2)}$ and Type-I phase-matching (PM) range of wavelengths (incident λ)

Compound	$\chi^{(2)}$ (pm/V)	Ref.	Type-I PM (μm)*	Ref.
Cu_2CdSnS_4	62	[153]	≥ 2.1	[153]
$AgGaSe_2$	66	[62]	≥ 3.1	[160]
$ZnGeP_2$	150	[62]	≥ 2.0	[62]
$AgGaS_2$	36	[77]	≥ 1.8	[163]
$LiInSe_2$	17, 22	[90,96]	≥ 2.0	[95]
Cu_2ZnSiS_4	15	[153]	≥ 1.7	[153]
Li_2CdGeS_4	51	[157]	≥ 1.5	[157]
$LiInS_2$	6.8–15	[85,88-90]	≥ 2.3	[90]

*The practical phase-matching range at long wavelengths is limited by the transparency window of each compound.

5. THIRD-ORDER OPTICAL NONLINEARITY

Figure 4(a) shows the series of THG spectra observed from $AgGaSe_2$ (d = 2–20 μm) plotted as a function of λ_{THG}. We confirmed that any contribution to THG by sum frequency generation of SHG and fundamental radiation ($3\omega = 2\omega + \omega$) is negligible; this cascade process is unlikely because random scattering by powders prevents coherent overlap of both beams. The THG signals are weak when λ_{THG} is shorter than the bandgap wavelength of $AgGaSe_2$ (λ_g = 730 nm at room temperature) because of efficient bandgap absorption of THG light as well as MPA of the fundamental beam.

However, THG becomes efficient when λ approaches the mid-IR regime. Although the limitation of our InGaAs detector (due to high dark counts and single-channel detection) prevented us from probing the THG response at 3300 nm, we predict that the THG efficiency saturates at the zero-frequency limit, which is typical and well known from first-principles calculations [164]. As in the SHG measurements, a dip near λ_{THG} = 700–800 nm arises from linear absorption of the fundamental beam by the container (capillary tubes). Note that the effect is common to the reference and all samples as shown in Figures 3 and 4, but this does not affect our estimation of the THG coefficients.

The THG counts from Cu_2CdSnS_4 (green) under the same experimental conditions are plotted in Figure 4(b) on a semi-logarithmic (semi-log) scale. The overall THG trend clearly indicates a case for narrow-gap materials where the THG signals are strongly absorbed by the compound in the visible range. Since the bandgap of Cu_2CdSnS_4 is about 0.92 eV (λ_g = 1350 nm), it is important to note that the experimental THG counts, even at λ_{THG} ~1066 nm, were likely underestimated due to bandgap absorption. This in turn implies that the THG efficiency could be much stronger at deeper mid-IR ranges. However, as explained in Sec. 7, the narrow-gap nature of Cu_2CdSnS_4 results in a low LDT, limiting its practical NLO applications.

THG responses from Cu_2ZnSiS_4 (pink) and Li_2CdGeS_4 (blue) are also plotted in Figure 4(b). Their wide bandgaps (E_g = 3.0–3.2 eV and 3.15 eV, respectively) cause rather "flat" THG dispersions over the observation range since the absorption effect in the visible range is relatively negligible. Because of differences in bandgap energies, as well as relevant linear absorption coefficients, the THG coefficients of the samples must be determined at longer wavelengths, i.e., in the static limit ($\lambda \rightarrow \infty$), in which both linear absorption of THG and MPA of fundamental are minimal.

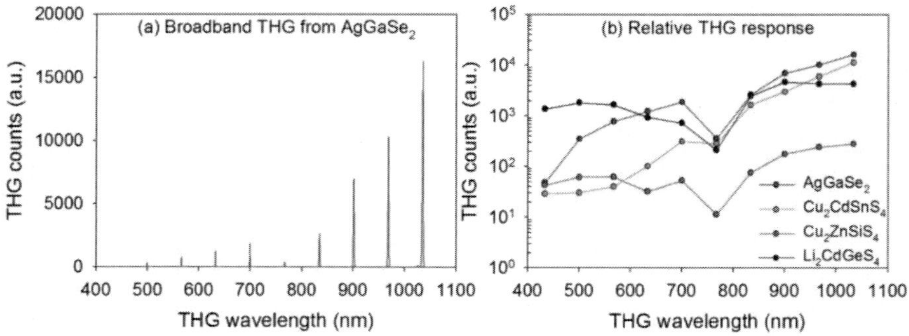

Figure 4. (Color online) (a) Wavelength-dependent THG spectra from $AgGaSe_2$. THG was very weak at λ =1100 nm, and therefore, not included. (b) Semi-log plot of relative THG counts from $AgGaSe_2$ (red), Cu_2CdSnS_4 (green), Cu_2ZnSiS_4 (pink) and Li_2CdGeS_4 (blue), respectively.

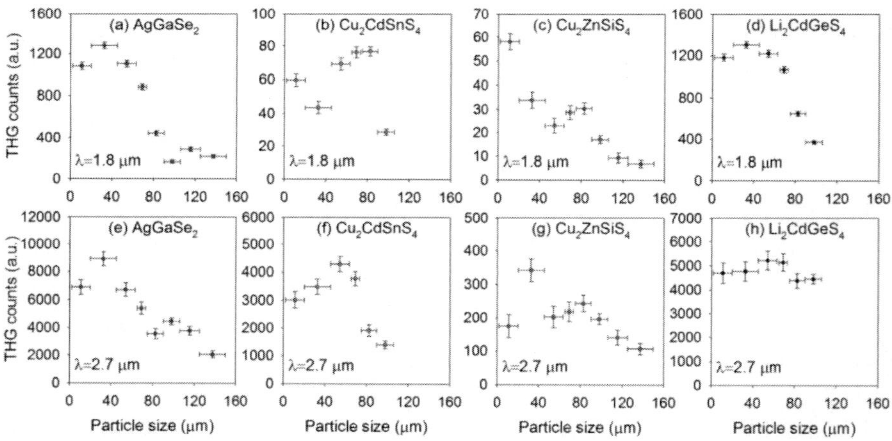

Figure 5. (Color online) First row: *d*-dependence of THG at λ = 1800 nm from (a) $AgGaSe_2$ (b) Cu_2CdSnS_4, (c) Cu_2ZnSiS_4 and (d) Li_2CdGeS_4. Second row: *d*-dependence of THG at λ = 2700 nm from (e) $AgGaSe_2$, (f) Cu_2CdSnS_4, (g) Cu_2ZnSiS_4 and (h) Li_2CdGeS_4.

Due to a large phase mismatch at λ and λ_{THG}, THG in many condensed matter systems is non-phase-matchable [165]. For instance, the decreasing trend in THG with increasing *d* for $AgGaSe_2$ at λ = 1800 nm and 2700 nm, in Figures 5(a) and 5(e), respectively, clearly indicates that THG is non-phase-matchable. The experimental THG coherence length is l_c = 32.5 ± 12.5 μm, exhibiting the maximum THG counts [16,20]. This value, together with the reference value of $\chi^{(3)} = 1.6 \times 10^5$ pm^2/V^2, was used for estimating $\chi^{(3)}$ values of the samples. Figures 5(b) and 5(f) show the case for Cu_2CdSnS_4. The particle size dependence at λ = 1800 nm is featureless since the phase-matching factor is washed out by strong bandgap absorption in the compound [166]. At λ = 2700 nm, however, it indicates a non-phase-matching case with a relatively long THG coherence length of l_c = 54 ± 9 μm. Note that the THG efficiency of Cu_2CdSnS_4 is significantly enhanced at 2700 nm compared to that at 1800 nm, but still affected by bandgap absorption as explained above.

Figures 5(c) and 5(g) show the THG particle size dependence of Cu_2ZnSiS_4 at these two input wavelengths, also indicating a non-phase-matching case. Note that the THG coherence length increases from 15 ± 5 μm at 1800 nm to 32.5 ± 12.5 μm at 2700 nm. This trend in a wide-bandgap material is typical, due to a reduced phase mismatch at longer wavelengths, which is also observed from Li_2CdGeS_4, as shown in Figures 5(d) and 5(h). Interestingly, Li_2CdGeS_4 exhibits almost phase-matched THG at $\lambda \geq$ 2700 nm, as shown in Figure 5(h).

The absolute $\chi^{(3)}$ value of each sample was estimated by comparing with the reference;

$$\chi_S^{(3)} = \chi_R^{(3)} \left[\frac{I_S^{THG}}{I_R^{THG}}\right]^{1/2} \frac{l_{c,R}}{l_{c,S}}, \tag{4}$$

where I_S^{THG} and I_R^{THG} are the measured THG counts from the sample and the reference at the same particle size and $l_{c,S}$ and $l_{c,R}$ are the corresponding THG coherence lengths determined from the data presented in Figure 5. Using $\chi_R^{(3)} = 1.6 \times 10^5$ pm^2/V^2, our calculation yields a $\chi^{(3)}(Cu_2CdSnS_4)$ ~ $(0.8 \pm 0.2) \times 10^5$ pm^2/V^2 at 3100 nm, but this is likely underestimated due to bandgap absorption as explained above. The estimated $\chi^{(3)}$ value for Cu_2ZnSiS_4 is about $(0.21 \pm 0.06) \times 10^5$ pm^2/V^2. This relatively low value is predicted due to its large bandgap [1]. Since Li_2CdGeS_4 seems to approach a phase-matching regime at $\lambda >$ 2700 nm, we estimated its $\chi^{(3)}$ value at 2700 nm, where both Li_2CdGeS_4 and $AgGaSe_2$ are non-phase-matchable with minimal THG absorption, yielding $\chi^{(3)}(Li_2CdGeS_4)$ ~ $(1.0 \pm 0.3) \times 10^5$ pm^2/V^2.

6. Bandgap and Optical Transparency

Diffuse reflectance UV-visible-NIR spectra were collected for Cu_2CdSnS_4, Cu_2ZnSiS_4 and Li_2CdGeS_4 polycrystalline powders obtained via solid-state synthesis [153,154,157] and used to estimate the optical bandgaps, Table 2. Using the Kubelka-Munk equation [167], reflectance was converted to absorption. Urbach energies were estimated for Cu_2CdSnS_4 and Li_2CdGeS_4 by fitting absorption vs. energy curves with $f(E) = A\exp[(E–E_g)/E_u]$, where A is a constant, E_g is the bandgap, E is the photon energy and E_u is the Urbach energy, so that the tailing effects could be considered for bandgap fitting. The data for Cu_2CdSnS_4 and Li_2CdGeS_4 were best fit using the function for an indirect bandgap, $\alpha = A(E–E_g)^{1/2}/E$, and a direct bandgap, $\alpha = A(E–E_g)^2/E$, respectively. Since α- and β-Cu_2ZnSiS_4 have been prepared only as a mixture and the resulting spectrum is complex, the bandgaps were roughly estimated by extrapolating the absorption edges to the baseline.

According to the optical-gap measurement, Cu_2CdSnS_4 has an indirect bandgap of 0.92 eV, although the calculated electronic structure shows that the lowest-energy transition occurs at a single point within the Brillouin zone (i.e., has a direct bandgap). Extrinsic broadening by significant Urbach tailing of a direct optical transition cannot be dismissed from the interpretation of the optical data. Further, there are numerous low-energy transitions within close proximity that can account for the experimental observation. Electronic structure calculations reveal that α-Cu_2ZnSiS_4 has a slightly narrower bandgap than β-Cu_2ZnSiS_4; thus, the lowest-energy absorption edge (~3.0 eV) in the collected spectrum is assigned to α-Cu_2ZnSiS_4, while the edge at ~3.2 eV corresponds to the bandgap of β-Cu_2ZnSiS_4. Based on our experiment, Li_2CdGeS_4 is a direct-gap semiconductor (E_g = 3.15 eV). Our calculations, for which the crystallographically determined unit-cell parameters were used without volume optimization, yielded a direct bandgap [157], in contrast to an indirect bandgap that was calculated by others using volume-optimized unit cell parameters [168].

The attenuated total reflectance (ATR) IR spectra were used in conjunction with the diffuse reflectance UV-visible-NIR spectra to determine the range of optical transparency for the three DLS systems. A Thermo Nicolet 380 FT-IR spectrometer equipped with an ATR accessory was used to collect the IR spectra, comprised of 64 scans, from 400 cm^{-1} to 4000 cm^{-1}. The OMNIC software was used for data collection and analysis. Since our FT-IR system uses a diamond crystal in optical contact with the sample, the depth of penetration into the sample is ~2 μm. That is near the limit of our smallest particle sizes. Thus, any thickness-dependent scaling effect on the intensity of the measured spectrum is negligible [169].

Cu_2CdSnS_4, Cu_2ZnSiS_4 and Li_2CdGeS_4 exhibit optical transparency into the far-IR. The bandgap absorption serves as the low-wavelength boundary of the clarity window. Within DLSs, bandgaps are easily tuned as a function of composition. Thus, the low-wavelength boundary of the clarity window is tuned from the border of the UV and visible regions (Cu_2ZnSiS_4 and Li_2CdGeS_4) to the near-IR (Cu_2CdSnS_4). Interestingly, these windows of transparency are wider than many of the commercially available materials, Table 2, and some of the newly discovered chalcogenides, e.g., $K_2P_2Se_6$, (0.596–19.8 μm) [131] and $Na_2Ge_2Se_5$ (0.521–18.2 μm) [170], with high nonlinearity in the IR. Since the clarity windows of Cu_2CdSnS_4, Cu_2ZnSiS_4 and Li_2CdGeS_4 extend into the far-IR, DFG is prone to extend into the THz region. However, practical DFG applications, as well as SHG applications, necessitate high LDTs.

7. Laser-Damage Threshold

MPA-induced damage determines the limit of useful performance of an NLO material. Therefore, an understanding of the mechanisms that cause radiation damage, in addition to measuring the LDT, is of great importance in the discovery of next-generation NLO materials working at long wavelengths. The green dots in Figure 6(a) correspond to the spectrally integrated SHG counts from Cu_2CdSnS_4 when the input intensity was varied within our experimental range. The solid line represents the maximum SHG case in which fundamental depletion is absent, i.e., $I_{SHG} = aI^2$, where I_{SHG} and I are the SHG and fundamental intensities with a being a proportionality constant that incorporates $|\chi^{(2)}|^2$. Although the compound has high optical nonlinearity, it undergoes serious damage upon laser illumination due to efficient 1-photon absorption (1PA) at 1064 nm. The LDT by 1PA seems to be significantly below 1 GW/cm^2 [175].

The pink dots in Figure 6(b) illustrate the case for Cu_2ZnSiS_4. Considering the bandgap of the compound, the deviation from the solid line (maximum SHG) at $I > 2$ GW/cm^2 arises from 3-photon absorption (3PA) of the fundamental beam. We estimated the corresponding 3PA coefficient, γ, by fitting the measured SHG power dependence using a modified fundamental intensity by 3PA, I_{3PA};

$$I_{\mathrm{SHG}} = aI_{\mathrm{3PA}}^2 \text{ with } I_{\mathrm{3PA}} = \frac{I}{\sqrt{1+2\gamma dI^2}}, \tag{5}$$

where $d = 90\text{–}106$ μm is roughly the particle size for our reflection-based collection geometry and the constant a was carefully determined by fitting the low-intensity regime where 3PA is absent. The red curve is a fit using Eq. (5), yielding $\gamma = 1.22 \pm 0.10$ cm^3/GW^2. This value is rather high in comparison to typical 3PA coefficients of inorganic materials having similar bandgaps [176,177], which may indicate that 3PA-induced optical damage is significant.

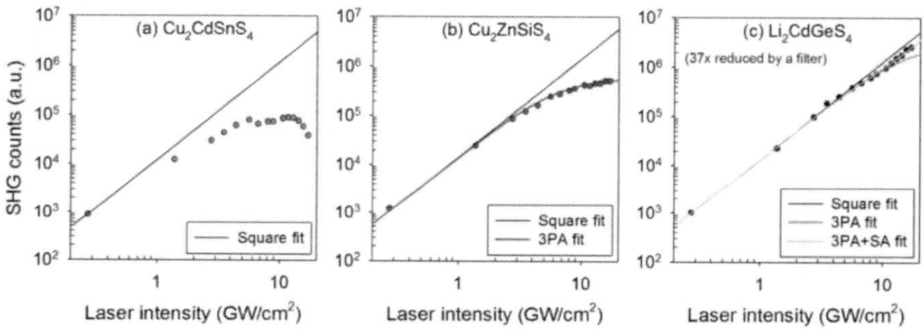

Figure 6. (Color online) SHG power dependence of (a) Cu_2CdSnS_4, (b) Cu_2ZnSiS_4 and (c) Li_2CdGeS_4 at λ = 1064 nm, superimposed by square fits (black), 3PA fits (red) and 3PA+SA fit (grey), respectively. Li_2CdGeS_4 has a very high LDT.

Table 2. Bandgaps and transparency windows.

Compound	Bandgap (eV/μm)	Ref.	Transparency Window (μm)	Transparency Level	Ref.
Cu_2CdSnS_4	0.92/1.35	[153]	1.37–25	60%	[153]
$AgGaSe_2$	1.83/0.68	[72]	0.71–19	"0" level	[171]
$ZnGeP_2$	2.0/0.62	[87]	0.74–12	"0" level	[172]
$AgGaS_2$	2.62/0.47	[72]	0.47–13	"0" level	[173]
$LiInSe_2$	2.86/0.43	[98]	0.72–10.4	$\alpha = 1$ cm^{-1}	[98]
α-Cu_2ZnSiS_4 β-Cu_2ZnSiS_4	~3.0/0.41 ~3.2/0.39	[154]	0.70–25	75%	[153]
Li_2CdGeS_4	3.15/0.39	[157]	0.50–23.5	75%	[174]
$LiInS_2$	3.6/0.34	[91]	0.34–13.2	"0" level	[91]

Table 3. LDT at $\lambda = 1.064$ μm and $\tau = 30$ ps, except for Cu_2CdSnS_4 where $\lambda = 1.3$ μm.

Compound	LDT (GW/cm^2)	Photon Absorption (PA) Process
Cu_2CdSnS_4	0.2	2PA; $\beta = 120$ cm/GW
$AgGaSe_2$	0.2	2PA; $\beta = 38$ cm/GW
$AgGaS_2$	0.1	2PA; $\beta = 40.5$ cm/GW
Cu_2ZnSiS_4	2.0	3PA; $\gamma = 1.22$ cm^2/GW^2
Li_2CdGeS_4	$> 5^*$	3PA; $\gamma = 0.22$ cm^2/GW^2 ($I_S = 14$ GW/cm^2)

*Li_2CdGeS_4 undergoes saturable absorption.

Since SHG from Li_2CdGeS_4 was strong enough to saturate our detector at 1064 nm, we used a neutral density filter to reduce the detection efficiency by 37× at the SHG conversion wavelength of 532 nm. We found that Li_2CdGeS_4 exhibits a rather interesting SHG power dependence (blue dots), Figure 6(c). The SHG power dependence cannot be explained by a single 3PA fit [see the 3PA fit (red curve)]. For $I < 5$ GW/cm^2, the SHG counts are well explained by $I_{SHG} = aI^2$. For 5 GW/cm^2 $< I <$ 10 GW/cm^2, a slight fundamental depletion by 3PA was observed. However, the compound shows saturable absorption (SA) as indicated by a square-law tendency for $I >$ 10 GW/cm^2: The slope is parallel to the solid line for the maximum SHG case. In order to fit the overall trend, we modified the 3PA coefficient by incorporating SA;

$$I_{\text{SHG}} = aI_{\text{3PA}}^2 \text{ with } I_{\text{3PA}} = \frac{I}{\sqrt{1+2\gamma dI^2/[1+(I/I_s)^3]}} \text{ with } \gamma \rightarrow \gamma/[1+(I/I_s)^3], \quad (6)$$

where I_s is a parameter known as the saturation intensity. For $I < I_s$, the power dependence is similar to Eq. (5), but it recovers the square-law behavior due to $I_{\text{3PA}} \propto I$ for $I > I_s$. The grey trace is a best fit using Eq. (6), yielding $\gamma = 0.22 \pm 0.02$ cm^3/GW^2 and $I_s = 14 \pm 2$ GW/cm^2. With this mild 3PA coefficient, a persisting square dependence unambiguously demonstrates that Li_2CdGeS_4 is particularly promising for high-power NLO applications.

8. Effect of Bandgap Energy on $\chi^{(2)}$ and LDT

It is very difficult to computationally predict $\chi^{(2)}$; thus it is best determined experimentally. One reason is that the $\chi^{(2)}$ is very sensitive to the magnitude of the bandgap and lattice constants, parameters that calculations are often inaccurate in predicting. Thus far, many of the calculations that have been performed are not particularly accurate in matching experiment [178]. Nevertheless, some progress has been made in recent years in the modeling of ABO_3 materials such as $LiNbO_3$ using first-principles calculations [179]. Also, density functional theory (DFT) calculations of several novel chalcogenides were recently carried out within a self-consistent formalism [164]. However, all DFT calculations of $\chi^{(2)}$ require *experimental input* (precise bandgap), so-called "rigid band shift", in order to correct the well-known bandgap underestimation. Here we take a rather phenomenological approach in order to correlate optical nonlinearity with bandgap.

Starting with the work of Miller in 1964 [180], researchers have tried to find correlations between NLO susceptibility and other materials properties. Miller's rule expressed $\chi^{(2)}$ in terms of dielectric constants. Similarly, Phillips and Van Vechten proposed the dielectric theory of electronegativity in covalent solids for the calculation of $\chi^{(2)}$ and $\chi^{(3)}$ values [181-183]. Others, such as Flytzanis and Ducuing [184] and Levine [185], extended this work with a continued focus on bonding. Flytzanis and Ducuing found correlation between $\chi^{(2)}$ and the dipole moment of the cation-anion bond. Levine's calculation of $\chi^{(2)}$ used bond ionicity, the difference between the tetrahedral covalent radii of the atoms in the bond and the *d*-electron contributions [186]. The relation between the index of refraction and bandgap was first reported by Moss in 1950 [187]. Subsequent work along these lines continues [188,189].

Over the years, it has become generally accepted that large bandgaps result in relatively higher LDTs and weaker NLO susceptibilities, while narrow bandgaps yield more impressive $\chi^{(2)}$ values and lower LDTs. Some researchers describe this as a "balance" and a key issue in the search for new materials [190,191].

Bandgap tuning is especially important in IR NLO materials since the LDT is reliant on factors that are closely related to bandgap absorption; wider bandgaps can give rise to higher LDTs. Incorporating elements that can result in compounds with wider bandgaps may provide access to improved SHG materials [129] with higher LDTs. For example, $Li_2Ga_2GeS_6$ (E_g = 3.65 eV) is more resistant to laser damage than the closely related Ag analogue, $AgGaGeS_4$ (E_g = 2.95 eV). The Li-containing material shows no degradation in SHG signal under prolonged laser irradiation, while the Ag analogue loses ~90% of its SHG efficiency after ~100 laser pulses [125]. In another example, the wide-gap semiconductor $LiInS_2$ (E_g = 3.56 eV) has an LDT that is an order of magnitude higher than $AgGaS_2$ (E_g = 2.73 eV) [192]. However, it is commonly accepted that high LDTs and high SHG coefficients are mutually exclusive and that a balance between LDT and $\chi^{(2)}$ must be achieved in order to acquire optimal SHG materials for use in the IR. As various compositions can be employed for bandgap tuning, our I_2-II-IV-VI_4 DLS system provides a valuable platform to study the relationships between electronic structures and NLO properties, including LDT.

Figure 7, which is an adaptation of a 1997 report by Jackson et al. [77], illustrates the strong correlation between wide bandgap and small $\chi^{(2)}$ as well as narrow bandgap and large $\chi^{(2)}$ using a variety of NLO materials that includes pnictides, chalcogenides, oxides, halides, etc. [77]. The benchmark materials $AgGaSe_2$, $AgGaS_2$, and $ZnGeP_2$ as well as the quaternary DLSs presented in this chapter, Cu_2CdSnS_4, Cu_2ZnSiS_4, and Li_2CdGeS_4, are highlighted.

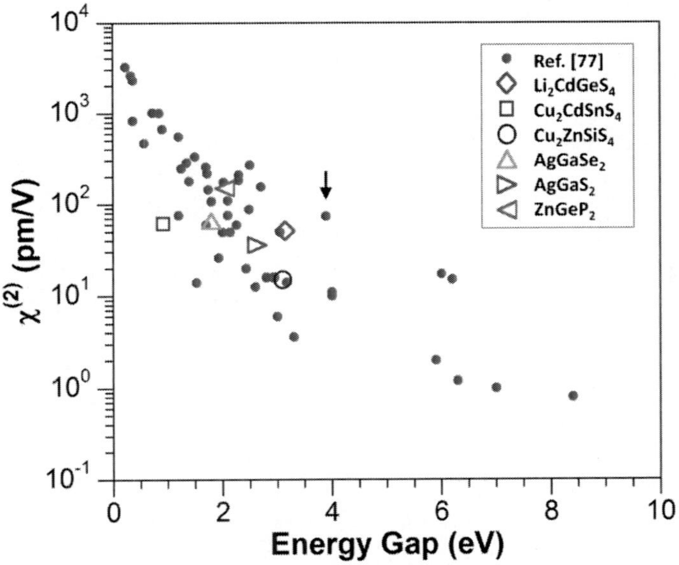

Figure 7. (Color online) The second-order NLO susceptibility ($\chi^{(2)}$) is plotted as a function of energy gap for selected compounds (•) from Ref. [77] in addition to Li_2CdGeS_4 (◇), Cu_2CdSnS_4 (□), α/β-Cu_2ZnSiS_4 (○), $AgGaSe_2$ (△), $ZnGeP_2$ (◁) and $AgGaS_2$ (▷). The arrow corresponds to ZnS. Adapted from Ref. [77] with permission. Copyright 1997 Elsevier.

Jackson et al. reported that two power-law expressions were required to fit $\chi^{(2)}$ versus E_g and figure of merit versus bandgap plots, one expression for the narrower bandgap region, $E_g < 1.2$ eV, and the other for the higher bandgap region of ~1.2–8.4 eV. One of the two empirical equations can be used to predict the $\chi^{(2)}$ value for a given compound using only the bandgap. Although this calculation does not provide a highly accurate prediction of $\chi^{(2)}$, the calculated $\chi^{(2)}$ values agreed with experimental values within a factor of two for 70% of the compounds, within a factor of three for 80% of the compounds and within a factor of four for 95% of the compounds. These results are unsurprising considering the 'scattered' data points. Deviation of the measured $\chi^{(2)}$ from the power-law prediction indicates strong or weak performance in comparison with the energy gap of the material. For example, $ZnGeP_2$ and GaSe are strong performers, with observed $\chi^{(2)}$ values being much larger than predicted. On the other hand, some compounds, deemed weak performers, have $\chi^{(2)}$ values that are lower than what one would expect based on bandgap values alone. For example, the calculated $\chi^{(2)}$ value for $CuInS_2$ is overestimated by a factor of ~11.

Cu_2CdSnS_4 has the highest $\chi^{(2)}$ value among the quaternary DLSs presented in this chapter and it seems impressive since it is comparable to the commercially available $AgGaSe_2$ and exceeds that of $AgGaS_2$. However, considering the bandgap energy of 0.92 eV, Cu_2CdSnS_4 is a weak performer; the calculated $\chi^{(2)}$ using Jackson's formula for $E_g < 1.2$ eV, $\chi^{(2)}$(pm/V) = $589/E_g$(eV) is 640 pm/V, while the actual $\chi^{(2)}$ was found to be 62 pm/V. Here the predicted $\chi^{(2)}$ value was overestimated by a factor of ~10. The calculated $\chi^{(2)}$ (pm/V) = $501/E_g^{2.7}$(eV) for wide-gap Cu_2ZnSiS_4 is also overestimated, but only by a factor of ~1.6, making Cu_2ZnSiS_4 a relatively average performer considering the bandgap of 3.0–3.2 eV. On the other hand, Li_2CdGeS_4 is an over achiever since the predicted $\chi^{(2)}$ is only ~23 pm/V while the measured value is 51 pm/V.

The wide-bandgap Li_2CdGeS_4 (E_g = 3.15 eV) seems to be very special since it is not only a high achiever in $\chi^{(2)}$ based on bandgap, but also the LDT is exceptional, Table 3. While many of the IR NLO chalcogenides have relatively narrow bandgaps, the widest gaps have been achieved through the incorporation of Li as discussed in Sec. 1.4. In terms of bandgap, $LiInSe_2$ (E_g = 2.86 eV) is the closest ternary Li-containing material to Li_2CdGeS_4. According to Jackson's formula, Li_2CdGeS_4 should have a lower $\chi^{(2)}$ as compared to $LiInSe_2$; however the measured $\chi^{(2)}$ = 51 pm/V of Li_2CdGeS_4 is more than double that of $LiInSe_2$. In fact, this value is higher than $\chi^{(2)}$ of all the ternary Li-containing materials, $LiAQ_2$ compounds (A = Al, In or Ga and Q = S or

Se), which are rather weak or average performers [85,88-90,96]. It is also interesting to compare Li_2CdGeS_4 and Cu_2ZnSiS_4, which share roughly the same bandgap energy; yet, the observed $\chi^{(2)}$ value of Li_2CdGeS_4 is more than three times that of Cu_2ZnSiS_4. As indicated by the arrow in Figure 7 and to the best of our knowledge, only ZnS with $E_g = 3.9$ eV has a higher $\chi^{(2)} \sim 74$ pm/V but the material is optically isotropic and therefore non-phase-matchable [87]. Therefore, the $\chi^{(2)}$ value of Li_2CdGeS_4 is the *highest* among phase-matchable NLO materials having bandgaps over 3.0 eV.

According to the data displayed in Figure 7, the $\chi^{(2)}$ value can vary over an order of magnitude for a given bandgap (e.g., $E_g = 2$ eV). These disparities arise because $\chi^{(2)}$ is not solely determined by bandgap, but also by other important materials properties. Parameters such as dielectric constant, bond ionicity, joint density of states and optical dipole matrix elements for the associated bands [164], etc., should be considered and optimized during the discovery phase of new materials for the achievement of high NLO susceptibility. As a part of our ongoing work with quaternary DLSs, additional compounds will be prepared and the optical nonlinearities evaluated. Placement of these data points in the bandgap-$\chi^{(2)}$ plot will help elucidate the property trend under compositional variation. We believe that this will generate a rationale for further enhancement in terms of $\chi^{(2)}$ and LDT in the future synthesis of advanced DLSs. For the present, we have tried to understand these phenomena by considering the basic bonding character of the compounds via electronic structure calculations.

9. Electronic Structure Calculations

The electronic structures, in particular the partial density of states (PDOS), for some of the benchmark materials are first reviewed here, in order to put the electronic structures of Cu_2CdSnS_4, Cu_2ZnSiS_4 and Li_2CdGeS_4 into perspective as detailed later.

The electronic structures of the chalcopyrite-type Cu-III-VI_2 and Ag-III-VI_2 compounds, where III = Al, In or Ga and VI = S, Se or Te, have been known for some time [193] and there is general agreement that the states at the valence band maximum, VB_{max}, are dominated by contributions mainly arising from Cu/Ag-*d* orbitals hybridizing with chalcogen-*p* orbitals. Almost all of the coinage metal *d*-character arises in the VB_{max}; there is negligible contribution of coinage metal-*d* states in the conduction band. This repulsive (coinage metal-*d*)-(chalcogen-*p*) interaction is stronger for sulfides than the corresponding selenides. The observed reduction in bandgap of the I-III-VI_2 compounds compared to the binary analogues, referred to as optical bowing [194], is mostly attributed to this strong *d-p* repulsion. For example, it has been noted that the bandgap energy of $AgGaSe_2$ is about 0.3 eV smaller than that of the binary analogue, $Zn_{0.5}Cd_{0.5}Se$ [195].

In the specific cases of $AgGaS_2$ and $AgGaSe_2$, this *d-p* repulsive interaction at the VB_{max} is slightly lesser than that found for the Cu analogues; but, it is still by far the dominant feature in the VB_{max}. For Cu-III-VI_2 compounds it has been estimated that the upper most valence bands are 40% *d*-like, whereas these states are 20% *d*-like in the case of the Ag analogues [196]. The PDOS plots for $AgGaS_2$ and $AgGaSe_2$ are qualitatively very similar with some small shifts in energies due to the differently sized chalcogen atoms. In the case of $AgInS_2$, deeper in the valence band, the Ga-*d* and S-*s* states are concentrated mainly between –16 eV and –12 eV, while the Ga-*s* states are localized around –6 eV to –8 eV. The contributions from the Ga-*p* states are diffused throughout the region of 0 eV to –5 eV, and do not make a major contribution to any one specific energy region. In these compounds, the states at the conduction band minimum, CB_{min}, arise chiefly from the Ga-*s* and S-*p* states with some contribution also from S-*s* states. There is no contribution from Ag-*d* states in the CB_{min} [197-199]. While the energies may differ slightly, the same general features have been calculated by others [200-202].

The electronic structures of the ternary chalcopyrite-type pnictides are different. In the case of $ZnGeP_2$ the VB_{max} is dominated by P-*p* states hybridized with Zn-*s*, Zn-*p* and Ge-*p* states. Below the VB_{max} is a band referred to as the IV-V bonding band. The Zn-*d* states are localized further down in the valence band and overlap by a small amount with the IV-V bonding band. In stark contrast to Cu/Ag-*d* states in the Cu-III-VI_2 and Ag-III-VI_2 compounds described above, the Zn-*d* states do not make any significant contributions to the VB_{max} or the CB_{min} [203].

Likewise, the bonding and optical properties of the Li-III-VI_2 materials are not affected by hybridization between cation-*d* and anion-*p* states [98]. The lack of this *d* orbital contribution near the Fermi level opens up the bandgap and increases the LDTs of these materials [204]. The use of Li also serves to increase the frequencies of crystal lattice vibrations and the Debye temperature; as a result, the thermal conductivities of the Li-containing materials are higher than those of the corresponding Ag analogues [205].

Electronic structure calculations were performed for Li-III-Se_2 (III = Al, Ga, or In) compounds with the β-$NaFeO_2$ structure [206]. The PDOS shows a trivial contribution of Li states in the vicinity of the Fermi level. The Li ions are basically just electron donors. As expected the Ga/In-*d* states are highly localized and deep in the valence band. The states at the VB_{max} are dominated by the Se-4*p* and the III-*p* contributions. The states at the CB_{min} are mostly attributed to the antibonding interaction of III-*s* and Se-4*p* orbitals. The variation in bandgap across the series is attributed to the difference in energy of the III-*s* orbitals. More recent calculations for $LiGaS_2$ show similar results [207]. Comparison of the Li-III-S_2, Li-III-Se_2 and Li-III-Te_2 electronic structures indicates that the electronic structures are more dependent upon the identity of the elements as opposed to the specific crystal structure.

The Li-III-Te_2 compounds do not possess the β-$NaFeO_2$ structure, but rather the same chalcopyrite structure as the Ag-containing ternaries. Electronic structure calculations for Li-III-Te_2 indicate that the VB_{max} is composed of mainly Te-*p* and III-*p* states, while the bottom of the conduction band arises primarily from the Te-*p* and III-*s* states with some contribution from the III-*p* states [208]. As in the case of the sulfides and selenides, the contribution of Li in the vicinity of the Fermi level is very minor. The Li-III-Te_2 electronic structures allow more direct comparison to the Ag-ternary compounds, since they have the same crystal structure. The widening of the bandgap observed for the Li compounds is most certainly a consequence of having Li in the compound instead of a coinage metal.

One relatively new NLO material discovered in 2010, $BaGa_4Se_7$, is reported as having an SHG response 2–3 times that of $AgGaS_2$ when irradiated with a 1064-nm laser [209]. The compound does not have a diamond-like structure; but, interestingly, the DOS for this compound shows similar features to those of the Li-III-VI_2 compounds discussed above. The $[Ga_4Se_7]^{2-}$ anionic framework defines the bandgap and the Ba cations act mainly as electron donors. The VB_{max} is dominated by Se-4*p* states as well as significant contribution from Ga-*p* states, while lower in the conduction band the states are mainly composed of Ga-*s*, Ga-*p*, Se-*s* and Se-*p* states. The Ba-*p* orbitals as well as the Ga-*d* orbitals lie very deep and localized in the valence band. The Ba-*p* states and the majority of the S-*s* states are also relatively deep in the valence band. The CB_{min} does have some donation from the Ba-*d* orbitals; however, not at the very most bottom region that defines the bandgap. Some recent progress has been made in the crystal growth of this material ($8 \times 8 \times 2$ mm^3); however, the compound has a very low thermal conductivity (< 1 W/mK), and the optical transparency region only extends to 14 μm. Yet, these early results are encouraging [210].

The electronic structures of Cu_2CdSnS_4, α-Cu_2ZnSiS_4, β-Cu_2ZnSiS_4 and Li_2CdGeS_4 were calculated using the total-energy code CASTEP, which uses

DFT and the plane-wave pseudopotential method to solve the one-electron Kohn-Sham equations. Computational details can be found elsewhere [153,157]. The total density of states (TDOS) and PDOS for Cu_2CdSnS_4, α-Cu_2ZnSiS_4 and Li_2CdGeS_4 are displayed in Figure 8. The overall DOS for β-Cu_2ZnSiS_4 are strikingly similar to those of α-Cu_2ZnSiS_4; there exist only subtle differences between the two that have been explained in earlier work [155].

In the case of Cu_2CdSnS_4, the VB_{max} is defined primarily by Cu-*d* and S-*p* states, similar to the VB_{max} of the coinage-metal-containing chalcopyrites, such as $AgGaS_2$ and $AgGaSe_2$. There are almost no contributions from the Cd and Sn orbitals at the VB_{max}. The Cd-*d* states are localized deep in the valence band around –9 eV to –10 eV, the Cd-*s* states are concentrated around –5 eV and the Cd-*p* states are diffused throughout the region of 0 eV to –5.5 eV, with no significant contribution in any one region. The Sn-*s* states, together with S-*s* and *p* states, are confined to a narrow region of energy around –7 eV, while the Sn-*p* states contribute mainly in the vicinity of –3 eV to –5.5 eV. The dominant contributions to the CB_{min} are from the Sn-*p* and S-*p* states; there is much weaker donation from Cu-*p*, Cd-*s* and Cd-*p* states.

While the bandgap of Cu_2ZnSiS_4 is over 2 eV wider than that of Cu_2CdSnS_4, the PDOS for the two compounds are qualitatively very similar in that the VB_{max} is dominated by the *d*-*p* hybridization of Cu and S, respectively. The Zn-*d* states are very deep in the valence band and highly localized, as are the states with mainly Cd-*d* character in Cu_2CdSnS_4. The Zn-*s* orbitals, although somewhat more diffuse and of slightly different energy, are relatively deep in the valence band and do not contribute near the VB_{max}, just as the Cd-*s* orbitals in Cu_2CdSnS_4. The contributions of the Si-*p* orbitals, like those of Sn-*p*, are spread out in the valence band and make their main donations at energies lower than –3 eV. The main difference in the PDOS of the two compounds lies in the composition of the CB_{min}. The widening of the bandgap for the Si compound is a result of the Si-*s* orbitals giving rise to states higher in the conduction band, whereas the Sn-*s* orbitals are lower in energy and a major contributor to the states near the CB_{min} defining the bandgap. This one variance results in significant differences in $\chi^{(2)}$ and LDT for the two compounds.

A striking feature of the PDOS for Li_2CdGeS_4 is the lack of any sizeable contribution from the Li orbitals in the valence band states, as is the case for the Li-III-VI_2 compounds discussed earlier. The contributions of the Li atoms are mainly spread out over the conduction band states. The states at the VB_{max} arise primarily from the S-*p* orbitals with little contribution from the Cd-5*p* orbitals. The Cd-5*s* as well as the Ge-4*p* orbitals give most of their contributions a bit lower in the PDOS between –2 eV to –4.5 eV. Lower in the valence band around –6 eV to –7 eV the Ge-4*s* together with S-*s* and *p* orbitals make a large contribution and there is a small contribution from the Cd-*s* and *d* orbitals. The Cd-*d* orbitals make a larger and relatively localized contribution deep within the valence band around –7 eV to –9 eV. The states at the CB_{min} have chiefly Ge-4*s* and S-3*p* character, while there are also small contributions from the Cd-5*s* and Cd-5*p* states. The bandgap in this case is defined by the charge transfer from the S-3*p* to the Ge-4*s* orbitals. It can be envisioned that by keeping the Li in the compound, but changing the identity of the II metal, the bandgap would change minimally and the other parameters such as phase-matching range, NLO susceptibility, thermal stability, etc., may be tuned.

Considering the electronic structure calculations of the ternary diamond-like materials, as well as those of the quaternaries presented in this chapter, it is possible to choose appropriate elements to target new materials with widened bandgaps for higher LDT. In a given I-III-VI_2 or I_2-II-IV-VI_4 compound the bandgap will increase according to Cu < Ag < Li, as well as Sn < Ge < Si, and Te < Se < S, while the other elements are held constant. Therefore, to target the widest-gap materials, one might prepare compounds such as Li_2-II-SiS_4. The choice of the I ion has to be Li; a similar (I-*d*)-(chalcogen-*p*) interaction has been noted at the VB_{max} of Ag_2-II-IV-VI_4 compounds [199,211]. However, to the best of our knowledge, there are no Li_2-II-SiS_4 compounds reported in the literature to date and it is anticipated that these phases may be hygroscopic and not stable under normal environmental conditions. A logical second choice for achieving wide bandgaps is the Li_2-II-GeS_4 series, which has been demonstrated in this chapter, as Li_2CdGeS_4 has a bandgap greater than 3 eV. A switch to the Se analog is likely to lower the bandgap and increase the NLO susceptibility at the cost of reducing the LDT. The magnitude of the effects on LDT should be experimentally investigated for optimizing the balance (trade-off).

For a given bandgap, the determination of the exact NLO susceptibility, or in Jackson's terms whether the compound will be a strong, average or weak performer, is still somewhat of a debate. However, it is generally accepted that the key is the dipole moment of the cation-anion bond, in which the transition upon electromagnetic perturbation depends on the joint DOS as well as optical dipole matrix elements from the perspective of quantum mechanics. Levine's simple bond-charge model, which considers the bond ionicity, the difference in the atomic radii and the *d* electron contributions, is a popular model for the calculation of second-order NLO susceptibility and there have also been some recent extensions/adaptations of this model [212,213].

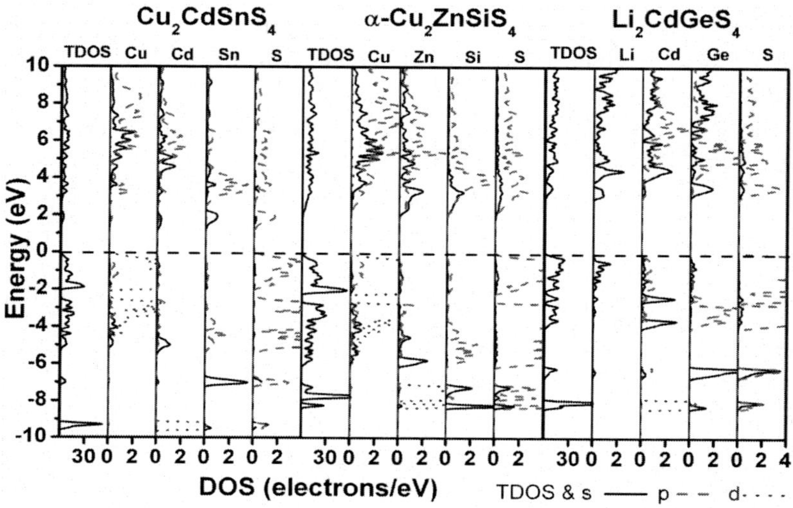

Figure 8. (Color online) TDOS and PDOS for Cu_2CdSnS_4, α-Cu_2ZnSiS_4 and Li_2CdGeS_4. The dotted black line denotes the Fermi level (E_F).

If ones compares Li_2CdGeS_4 to Cu_2ZnSiS_4, two compounds with nearly identical bandgaps, the difference in the observed $\chi^{(2)}$ values can be attributed to simple ideas from Levine's work. One of Levine's favorable bond properties, which leads to larger $\chi^{(2)}$, is increased covalency [185]. Our electronic structure calculations include the Mulliken bond population analysis for each crystallographically unique bond in the structure, where a value of 1 represents a purely covalent bond and a value of 0 a purely ionic one. The Cd-S and Ge-S bonds in Li_2CdGeS_4 are calculated to have greater covalency than the Zn-S and Si-S bonds in Cu_2ZnSiS_4, Table 4. This is most likely due to the presence of Li that acts nearly completely as an electron pair donor. The Cu cation in Cu_2ZnSiS_4 participates to a greater extent in covalent bonding with the S anion than the Li cation in Li_2CdGeS_4, thus keeping S from bonding more strongly with Zn and Si cations. Therefore, Li serves not only to open the bandgap in Li_2CdGeS_4, but to also allow the framework to be more covalent, thereby increasing the excess charge in the Ge-S and Cd-S bonding regions and ultimately $\chi^{(2)}$. Similarly in the case of $LiNbO_3$ and $LiTaO_3$, it is assumed that, since the Li-O bond has very low covalency, all of the susceptibility resides in the transition metal-oxygen bonds [185].

Table 4. Mulliken bond population data determined by electronic structure calculations using the CASTEP code

	Bonds				
Compound	Cu-S Avg. Bond Order	Zn-S Avg. Bond Order	Cd-S Avg. Bond Order	Si-S Avg. Bond Order	Sn-S Avg. Bond Order
Cu_2CdSnS_4	0.37	NA	0.32	NA	0.49
α-Cu_2ZnSiS_4	0.36	0.42	NA	0.70	NA
β-Cu_2ZnSiS_4	0.37	0.42	NA	0.69	NA
	Li-S Avg. Bond Order			Ge-S Avg. Bond Order	
Li_2CdGeS_4	0.11		0.49	0.75	

Longer bond distances within covalent bonds, created by larger sized ions bound together, is another of Levine's favorable bond properties for higher $\chi^{(2)}$ that serves to increase the region of charge between the atoms in the bond [185]. In agreement with this idea, the average bond distances in the $[CdGeS_4]^{2-}$ framework of Li_2CdGeS_4 are longer than those found in the $[ZnSiS_4]^{2-}$ framework of Cu_2ZnSiS_4.

A combination of our electronic structure calculations and Levine' bond-charge model can be used to target compositions that are likely to possess enhanced properties. For example, Ag_2ZnSiS_4 displays an optical bandgap of 3.28 eV [199], which is similar to that of Cu_2ZnSiS_4. Here the longer Ag-S bond distances should allow the $[ZnSiS_4]^{2-}$ framework to achieve greater covalency, thus the $\chi^{(2)}$ value is expected to be larger than that of Cu_2ZnSiS_4. Alternatively, the use of a larger II ion, such as Hg^{2+}, would produce longer II-S bond distances and serve to enhance $\chi^{(2)}$. Electronic structure calculations for these different element combinations should help to focus synthetic efforts.

10. Thermal Stability and Prospects for Future Crystal Growth

While small single crystals of tenths of a millimeter on an edge have been produced and used in the structure determination of Cu_2CdSnS_4 [153], α-Cu_2ZnSiS_4 [154] and Li_2CdGeS_4 [156], the preparation of large single crystals, greater than 1 cm^3, is of utmost importance for NLO applications. Polycrystalline reaction products, which cannot be prepared as pure phases, or meta-stable or kinetically stable compounds that can *only* be prepared in a molten flux are not usually amendable to large crystal growth by conventional means. The prospects of single-crystal growth will be better if the NLO material melts congruently. To assess congruent melting, X-ray powder diffraction (XRPD) of the thermal analysis residue must be used to confirm that the structure remains intact and that the sample is void of degradation or reaction products. The congruency of melting cannot be determined based solely on differential thermal analysis (DTA) or differential scanning calorimetry scans. In fact, $K_3Ta_2AsS_{11}$ and $Rb_3Ta_2AsS_{11}$ were deemed incongruent melters based on XRPD data, despite the fact that the thermal analysis diagrams contained only one endothermic and one exothermic peak [134]. Only if the XPRD patterns reveal the presence of a single-phase material before and after melting, can the NLO material be deemed as a congruent melter.

The congruent melting of Cu_2CdSnS_4 at 930 °C, as confirmed by XRPD patterns of the sample before and after the DTA experiment, implies great prospect of growing Cu_2CdSnS_4 as sizeable single crystals, Figure 9. Cu_2ZnSiS_4 exhibits higher thermal stability; a melting point was not observed in DTA scans up to 1000 °C. Li_2CdGeS_4 has the lowest melting point at ~890 °C.

Upon heating, the major endothermic event observed at 892 °C in the DTA curve of Li_2CdGeS_4 is expected to correspond to the melting point. A small endothermic peak appears at approximately 274 °C, and might be attributed to the melting of a minimal amount of residual Li_2S_x flux from the synthesis. Upon cooling, extra exothermic peaks appear and are likely caused by a reaction with the fused-silica ampoule, as evidenced from the compromised integrity of the ampoule per visual inspection. A small exothermic peak appears at approximately 270 °C and expectedly arises from the residual flux. The melting temperature of 892 °C corresponds with a reasonable thermal stability, and it is lower than some traditional NLO materials, e.g., $ZnGeP_2$, 1025 °C [214] and $AgGaS_2$, 997 °C [72]. We were unable to assess congruent melting from the DTA experiment because the compound attacks the ampoule. Alternatively, we conducted a similar experiment in a furnace in which the sample was placed inside of a graphite tube, inside of a fused-silica tube, and heated to 900 °C for 12 h. An XRPD pattern of the Li_2CdGeS_4 sample used for this experiment seemed to indicate a phase-pure material, however, the Li_2S_x residual flux may be amorphous. After cooling, the XRPD pattern of the product showed some amount of Li_2CdGeS_4 remaining, as well as a significant amount of CdS. These results could indicate that the compound does not melt congruently. However, we are hesitant to make that conclusion because the compound was not heated up alone; there was also presumably Li_2S_x present that could alter the result. Further work needs to be conducted in order to obtain Li_2CdGeS_4 as a pure-phase material.

There are some issues with Li when it comes to crystal growth because Li is known to react with fused-silica. In our reactions, we used graphite tubes inside of fused-silica and this is not impractical. Large, 10 mm in diameter and 20 mm in length, crystals of $LiInSe_2$ have been grown via the vertical Bridgman method using polished graphite crucibles inside of fused-silica tubes [98]. Although some Se loss was noted, a post-growth treatment in Se vapor with heating near the melting point was used to improve the crystal quality by removing opaque inclusions. The importance of high-quality single crystals cannot be minimized. The bandgap energy of $LiInSe_2$ is strongly dependent on growth conditions and values have been reported from 1.1 to 2.87 eV [143].

$LiInS_2$ suffers similar challenges as $LiInSe_2$, resulting in colorless, yellow, grey-rose and grey crystals depending on growth conditions [91]. Some of the color differences have been ascribed to Li deficiency and In excess [85,91]. The nonstoichiometry and consequential defects have contributed to small differences in the NLO properties of the different samples [85]. Some samples were found to be deficient in S; yet, this deficiency could be improved by post-growth annealing in S vapor [91]. Others have had success conducting the annealing step in Li_2S vapor at a temperature close to the melting point [82].

Due to issues mostly with anisotropic thermal expansion, the chalcopyrite crystals have not been easy to grow. Over the years, a number of methods have been developed to access crack-free, low-loss single crystals. A number of these methods use a seed crystal in the growth process. For example, the HGF-growth technique has been used to grow optical-quality $ZnGeP_2$ and $CdGeAs_2$ via oriented seeds [215]. Post-growth annealing is also necessary to ensure minimal absorption losses.

Similar techniques as those used to prepare single crystals of the ternary materials, such as vertical Bridgman or HGF, may be applied to the quaternary compounds discussed in this chapter. Post-growth annealing steps may also be necessary, as described above for the mature NLO materials.

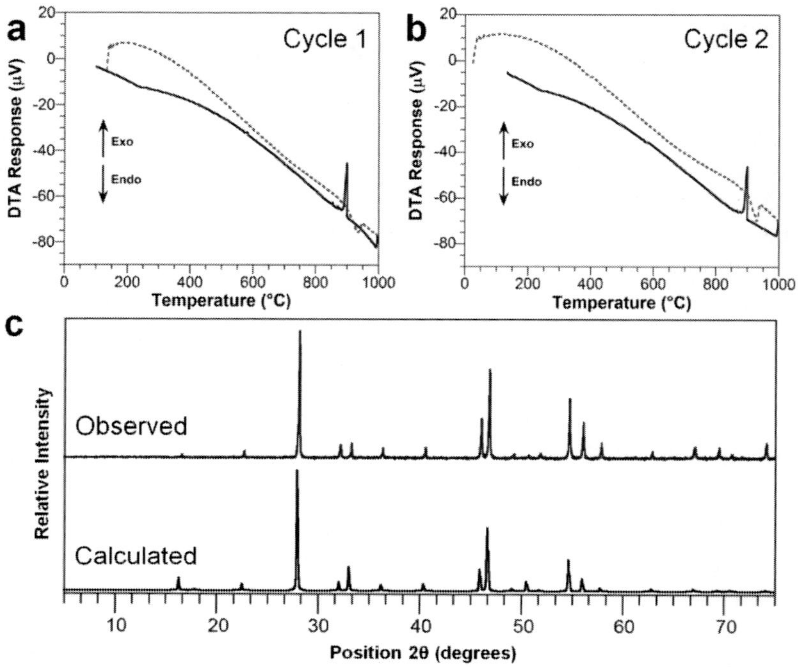

Figure 9. (Color online) (a) First- and (b) second-DTA cycle diagrams for Cu_2CdSnS_4. (c) XPRD pattern of the Cu_2CdSnS_4 DTA residue after the second-cycle compared to the calculated pattern.

CONCLUSION

In summary, the principles of nonlinear optics were presented to provide the basis of the NLO experiments described in this chapter. NLO materials for use with IR frequencies benefit a wide array of applications for military, medical, commercial and industrial sectors. However, NLO materials must be improved for many applications to become viable; thus, a number of criteria, in addition to NLO susceptibility, must be considered in the discovery of new NLO materials. Current benchmark materials were reviewed along with the advantages and disadvantages in regards to the ideal criteria. Several design strategies for discovering new IR NLO materials, as well as the adherence of the new materials to the criteria were assessed. The design strategy for the preparation of quaternary DLSs in pursuit of improved materials for IR NLO applications has been highlighted as promising.

This chapter presented the second- and third-order NLO properties of the I_2-II-IV-VI_4 DLSs, namely, the narrow-gap Cu_2CdSnS_4 and the wide-gap Cu_2ZnSiS_4 and Li_2CdGeS_4 and an assessment of these properties in regards to the criteria for ideal NLO materials. Broadband SHG measurements reveal wide ranges of phase-matchability. These three DLSs exhibit wide regions of optical transparency in the IR. The LDTs of the three DLS are ranked as Cu_2CdSnS_4 < Cu_2ZnSiS_4 < Li_2CdGeS_4. Interestingly, Li_2CdGeS_4 exhibits an exceptional LDT > 5 GW/cm^2 with saturable absorption persisting up to at least 14 GW/cm^2. Measurement past this point is a current experimental limitation due to melting of the glass tube. While wide bandgaps are generally ideal for NLO materials since they are associated with high LDT, they are also often correlated with low $\chi^{(2)}$. When categorized based on bandgap, according to Jackson's formula, Cu_2CdSnS_4 is considered to be a weak performer and Cu_2ZnSiS_4 is an average performer, while Li_2CdGeS_4 is a strong performer in terms of SHG. In fact, the phase-matched SHG efficiency of Li_2CdGeS_4 is the highest among wide-gap NLO materials with bandgaps larger than 3.0 eV. Though, we believe that further property optimization in this compound is viable through compositional derivatization.

Properties were rationalized and target materials have been proposed based on the results of electronic structure calculations and simple concepts from Levine's bond-charge model for the calculation of nonlinear susceptibility. For the full potential of these DLSs to be realized for applications, sizeable crystals should be grown; accordingly, the congruent melting of Cu_2CdSnS_4 implies great prospect.

Considering the three DLSs, Li_2CdGeS_4 holds the greatest potential, as it exhibits remarkable properties that are vital for laser applications. Comparable to $AgGaSe_2$, Li_2CdGeS_4 exhibits highly efficient phase-matched SHG conversion in the IR. More importantly, our SHG power-dependence experiments at the widely used Nd:YAG line indicate that Li_2CdGeS_4 greatly surpasses $AgGaSe_2$ in the context of high-power DFG for coherent generation of IR and THz frequencies owing to its wider range of optical transparency and higher LDT with no apparent 2PA, as well as higher thermal stability.

ACKNOWLEDGMENTS

This work was supported by the National Science Foundation, DMR-1201729. Use of the Advanced Photon Source, an Office of Science User Facility operated by the U.S. Department of Energy (DOE) Office of Science by Argonne National Laboratory, was supported by the U.S. DOE under Contract No. DE-AC02-05CH11357. Y.S.K. acknowledges support from the Basic Science Research Program (2014-010369) and Priority Research Center Program (2009-0093818) through the NRF of Korea funded by the Ministry of Education. We graciously thank Gary C. Catella (VP of Technology) and Carl Brunetta (Staff Scientist) from Gooch & Housego (Ohio) for fruitful discussions.

REFERENCES

[1] Boyd, RW. *Nonlinear Optics, Third Edition;* Academic Press: San Diego, CA, 2008.

[2] Paschotta, R; Fiedler, K; Kürz, P; Mlynek, J. "Nonlinear mode coupling in doubly resonant frequency doublers." *Appl. Phys. B*, 1994, 58, 117–119.

[3] Paschotta, R; Kürz, P; Henking, R; Schiller, S; Mlynek, J. "82% efficient continuous-wave frequency doubling of 1.06 μm with a monolithic MgO:LiNbO3 resonator." *Opt. Lett.*, 1994, 19, 1325–1327.

[4] Webjorn, J; Siala, S; Nam, DW; Waarts, RG; Lang, RJ. "Visible laser sources based on frequency doubling in nonlinear waveguides." *IEEE J. Quantum Electron.*, 1997, 33, 1673–1686.

[5] Durfee, CG; Rundquist, AR; Backus, S; Herne, C; Murnane, MM; Kapteyn, HC. "Phase matching of high-order harmonics in hollow waveguides." *Phys. Rev. Lett.*, 1999, 83, 2187–2190.

[6] Meyn, J.-P; Fejer, MM. "Tunable ultraviolet radiation by second harmonic generation in periodically poled lithium tantalate." *Opt. Lett.*, 1997, 22, 1214–1216.

[7] Hoyt, CW; Sheik-Bahae, M; Ebrahimzadeh, M. "High-power picosecond optical parametric oscillator based on periodically poled lithium niobate." *Opt. Lett.*, 2002, 27, 1543–1545.

[8] Canagasabey, A; Corbari, C; Gladyshev, AV; Liegeois, F; Guillemet, S; Hernandez, Y; Yashkov, MV; Kosolapov, A; Dianov, EM; Ibsen, M; Kazansky, PG. "High-average-power second harmonic generation from periodically poled silica fibers." *Opt. Lett.*, 2009, 34, 2483–2485.

[9] Fiore, A; Berger, V; Rosencher, E; Bravetti, P; Nagle, J. "Phase matching using an isotropic nonlinear optical material." *Nature*, 1998, 391, 463–466.

[10] Shi, W; Ding, YJ; Fernelius, N; Vodopyanov, K. "Efficient, tunable, and coherent 0.18-5.27-THz source based on GaSe crystal." *Opt. Lett.*, 2002, 27, 1454–1456.

[11] Cotter, D; Manning, RJ; Blow, KJ; Ellis, AD; Kelly, AE; Nesset, D; Phillips, ID; Poustie, AJ; Rogers, DC. "Nonlinear optics for high-speed digital information processing." *Science*, 1999, 286, 1523–1528.

[12] Barad, Y; Eisenberg, H; Horowitz, M; Silberberg Y. "Nonlinear scanning laser microscopy by third harmonic generation." *Appl. Phys. Lett.*, 1997, 70, 922–924.

[13] Tsang, T; Krumbugel, MA; DeLong, KW; Fittinghoff, DN; Trebino, R. "Frequency-resolved optical-gating measurements of ultrafast pulses using surface third-harmonic generation." *Opt. Lett.*, 1996, 21, 1381–1383.

[14] Sheik-Bahae, M; Said, AA; Wei, T; Hagan, DJ; Van Stryland, EW. "Sensitive measurement of optical nonlinearities using a single beam." *IEEE J. Quantum Electron.*, 1991, 26, 760–769.

[15] Sheik-Bahae, M; Hutchings, DC; Hagan, DJ; VanStryland, EW. "Dispersion of bound electronic nonlinear refraction in solids." *IEEE J. Quantum Electron.*, 1991, 27, 1296–1309.

[16] Jang, JI; Haynes, AS; Saouma, FO; Otieno, CO; Kanatzidis, MG. "Broadband studies of the strong mid-infrared nonlinear optical responses of $KPSe_6$." *Opt. Mater. Express*, 2013, 3, 1302–1312 – Special Issue: Mid-IR photonic materials.

[17] Mani, S; Jang, JI; Ketterson, JB. "Large third-order susceptibility and third-harmonic generation in centrosymmetric Cu_2O crystal." *Opt. Lett.*, 2009, 34, 2817–2819.

[18] Mani, S; Jang, JI; Ketterson, JB. "Nonlinear optical processes at quadrupole polariton resonance in Cu_2O as probed by a Z-scan technique." *Phys. Rev. B*, 2010, 82, 113203.

[19] Maker, PD; Terhune, RW; Nisenoff, M; Savage, CM. "Effects of dispersion and focusing on the production of optical harmonics." *Phys. Rev. Lett.*, 1962, *8*, 21–23.

[20] Kurtz, SK; Perry, TT. "A powder technique for the evaluation of nonlinear optical materials." *J. Appl. Phys.*, 1968, 39, 3798–3813.

[21] Gibbs, HM. *Optical Bistability*; Academic Press: New York, NY, 1985.

[22] Agrawal, GP. *Nonlinear Fiber Optics*; Academic Press: New York, NY, 2001.

[23] Gaeta, AL. "Collapsing light really shines." *Science*, 2003, 301, 54–55.

[24] Braun, A; Korn, G; Liu, X; Du, D; Squier, J; Mourou, G. "Self-channeling of high-peak-power femtosecond laser pulses in air." *Opt. Lett.*, 1995, 20, 73–75.

[25] Kasparian J; Wolf, J.-P. "Physics and applications of atmospheric nonlinear optics and filamentation." *Opt. Express*, 2008, 16, 466–493.

[26] Adams, DE; Planchon, TA; Hrin, A; Squier, JA; Durfee, CG. "Characterization of coupled nonlinear spatiospectral phase following an ultrafast self-focusing interaction." *Opt. Lett.*, 2009, 34, 1294–1296.

[27] Shim, B; Schrauth, SE; Gaeta, AL. "Filamentation in air with ultrashort mid-infrared pulses." *Opt. Express*, 2011, 19, 9118–9126.

[28] Denk, W; Strickler, JH; Webb, WW. "Two-photon laser scanning fluorescence microscopy." *Science*, 1990, 248, 73–76.

[29] Masters, BR; So, PT; Gratton, E. "Multiphoton excitation fluorescence microscopy and spectroscopy of in vivo human skin." *Biophys. J.*, 1997, 72, 2405–2412.

[30] Diaspro, A. "Developments in multiphoton excitation microscopy." *Microsc. Anal.*, 2012, 25th Anniversary Issue September, 51–56.

[31] Sheik-Bahae, M; Said, AA; Hagan, DJ; Soileau, MJ; Van Stryland, EW. "Nonlinear refraction and optical limiting in thick media." *Opt. Eng.*, 1991, 30, 1228–1235.

[32] Scalora, M; Dowling, JP; Bowden, CM; Bloemer, MJ. "Optical limiting and switching of ultrashort pulses in nonlinear photonic band gap materials." *Phys. Rev. Lett.*, 1994, 73, 1368–1371.

[33] Wang J; Blau, WJ. "Inorganic and hybrid nanostructures for optical limiting." *J. Opt. A: Pure Appl. Opt.*, 2009, 11, 024001.

[34] Hopkins, FK. "Military laser applications." *Opt. Photonics News*, 1998, 9, 32–38.

[35] Hecht, J. "Photonic frontiers: laser countermeasures: Scaling down mid-IR laser countermeasures for smaller aircraft." *Laser Focus World*, 2014, 50, 31–38.

[36] Lu, W; Liu, L; Sun, J; Pan, W. "Multi-channel 2-D pattern transfer model based on imaging optics for submarine laser uplink communication." *Opik*, 2008, 119, 388–394.

[37] Williams, OM. "Dynamic infrared scene projection: A review." *Infrared Phys. Techn.*, 1998, 39, 473–486.

[38] Richardson, MA; Coath, JA. "Infrared optical modulators for missile testing." *Opt. Laser. Technol.*, 1998, 30, 137–140.

[39] Eckardt, RC. "Scene projection by nonlinear optics." *Proc. SPIE*, 1995, 2469, 132–140.

[40] Knuteson, DJ; Singh, NB; Kanner, G; Berghmans, A; Wagner, B; Kahler, D; McLaughlin, S; Suhre, D; Gottlieb, M. "Quaternary $AgGaGe_nSe_{2(n+1)}$ crystals for NLO applications." *J Cryst. Growth*, 2010, 312, 1114–1117.

[41] Miller, MJ; Mott, AG; Ketchel, BP. "General optical limiting requirements." *Proc. SPIE*, 1998, 3472, 24–29.

[42] Zuclich, JA; Lund, DJ; Stuck, BE. "Wavelength dependence of ocular damage thresholds in the near-ir to far-ir transition region: proposed revisions to MPES." *Health Phys.*, 2007, 92, 15–23.

[43] Clewes, RJ; Howle, CR; Stothard, DJM; Dunn, MH; Robertson, G; Miller, W; Malcolm, G; Maker, G; Cox, R; Williams, B; Russell, M. "Stand-off spectroscopy for the detection of chemical warfare agents." *Proc. SPIE*, 2012, 8456, 85460X.

[44] Todd, MW; Provencal, RA; Owano, TG; Paldus, BA; Kachanov, A; Vodopyanov, KL; Hunter, M; Coy, SL; Steinfeld, JI; Arnold, JT. "Application of mid-infrared cavity-ringdown spectroscopy to trace explosives vapor detection using a broadly tunable (6-8 μm) optical parametric oscillator." *Appl. Phys. B*, 2002, 75, 367–376.

[45] Wang, C; Sahay, P. "Breath analysis using laser spectroscopic techniques: Breath biomarkers, spectra fingerprints, and detection limits." *Sensors*, 2009, 9, 8230–8262.

[46] Persijn, S; Harren, F; van der Veen, A. "Quantitative gas measurements using a versatile OPO-based cavity ringdown spectrometer and comparison with spectroscopic databases." *Appl. Phys. B*, 2010, 100, 383–390.

[47] Chatterjee, U. "Development of coherent tunable source in 2-15 μm region using nonlinear frequency mixing processes." *Pramana - J. Phys.*, 2014, 82, 29–38.

[48] Ferrante do Amaral, CE; Wolf, B. "Current development in non-invasive glucose monitoring." *Med. Eng. Phys.*, 2008, 30, 541–549.

[49] Tura, A; Maran, A; Pacini, G. "Non-invasive glucose monitoring: Assessment of technologies and devices according to quantitative criteria." *Diabetes Res. Clin. Pr.*, 2007, 77, 16–40.

[50] Dietz, N; Madarasz, FL. "Chemical and biological sensors based on optically confined birefringent chalcopyrite heterostructures." *Mat. Sci. Eng B-Solid*, 2003, 97, 182–195.

[51] Stoeppler, G; Shellhorn, M; Eichhorn, M. "Enhanced beam quality for medical applications at 6.45 μm by using a RISTRA ZGP OPO." *Laser Phys.*, 2012, 22, 1095–1098.

[52] Haidar, S; Miyamoto, K; Ito, H. "Generation of tunable mid-IR (5.5-9.3 μm) from a 2-μm pumped $ZnGeP_2$ optical parametric oscillator." *Opt. Commun.*, 2004, 241, 173–178.

[53] Andreev, Yu. M; Geiko, PP; Krekov, GM. "Detection of trace concentration of some simple pollutants in Tomsk." *Proc. SPIE*, 1991, 1811, 367–370.

[54] Vaicikauskas, V; Kaucikas, M; Swedas, V; Kuprionis, Z. "Mobile spectroscopic system for trace gas detection using a tunable mid-IR laser." *Rev. Sci. Instrum.*, 2007, 78, 023106.

[55] Sumpf, B; Rehle, D; Kelz, T; Dronfeldt, H.-D. "A tunable diode-laser spectrometer for the MIR region near 7.2 μm applying difference-frequency generation in $AgGaS_2$." *Appl. Phys. B.*, 1998, 67, 369–373.

[56] Bamford, DJ; Cook, DJ; Sharpe, SJ; Van Pelt, AD. "Widely tunable rapid-scanning mid-infrared laser spectrometry for industrial gas process stream analysis." *Appl. Opt.*, 2007, 46, 3958–3968.

[57] Kelemen, MT; Weber, J; Rattunde, M; Pfahler, C; Kaufel, G; Moritz, R; Manz, C; Mikulla, M; Wagner, J. "High-power diode laser arrays at 2 μm for materials processing." Lasers in manufacturing, 2005: proceedings of the third international WLT - Conference on lasers in manufacturing, Munich, Germany, June, 2005; AT-Fachverlag, Stuttgart, Germany, 2005, 713–718.

[58] Udem, Th; Holzwarth, R; Hänsch, TW. "Optical frequency metrology." *Nature*, 2002, 416, 233–237.

[59] Kirchengast, G; Schweitzer, S. "Climate benchmark profiling of greenhouse gases and thermodynamic structure and wind from space." *Geophys. Res. Lett.*, 2011, 38, L13701.

[60] Ding, YJ; Shi, W. "Observation of THz to near-infrared parametric conversion in $ZnGeP_2$ crystal." *Opt. Express*, 2006, 14, 8311–8316.

[61] www.inradoptics.com.

[62] Ohmer, MC; Pandey, R. "Emergence of chalcopyrites as nonlinear optical materials." *MRS Bull.*, 1998, 23, 16–22.

[63] Schunemann, PG. "Improved NLO crystals for mid-IR laser applications." *Proc. SPIE*, 2007, 6455, 64550R.

[64] Vodopyanov, KL; Schunemann, PG. "Efficient difference-frequency generation of 7-20 μm radiation in $CdGeAs_2$." *Opt. Lett.*, 1998, 23, 1096–1098.

[65] Schunemann, PG; Pollak, TM. "Single crystal growth of large, crack-free $CdGeAs_2$." *J. Cryst. Growth*, 1997, 174, 272–277.

[66] Vodopyanov, KL; Mirow, SB; Voevodin, VG; Schunemann, PG. "Two-photon absorption in GaSe and $CdGeAs_2$." *Opt. Commun.*, 1998, 155, 47–50.

[67] Mason, PD; Jackson, DJ; Gorton, EK. "CO_2 laser frequency doubling in $ZnGeP_2$." *Opt. Commun.*, 1994, 110, 163–166.

[68] Zawilski, KT; Schunemann, PG; Setzler, SD; Pollak, TM. "Large aperture single crystal $ZnGeP_2$ for high-energy applications." *J. Cryst. Growth*, 2008, 310, 1891–1896.

[69] Lei, Z; Zhu, C; Xu, C; Yao, B; Yang, C. "Growth of crack-free $ZnGeP_2$ large single crystals for high-power mid-infrared OPO applications." *J. Cryst. Growth*, 2014, 389, 23–29.

[70] Zotova, IB; Ding, YJ. "Optical parametric oscillators in the presence of strong two-photon absorption for extended applications of nonlinear optical materials." *Opt. Commun.*, 2001, 198, 453–458.

[71] Feigelson, RS; Route, RK. "Recent developments in the growth of chalcopyrite crystals for nonlinear infrared applications." *Opt. Eng.*, 1987, 26, 113–119.

[72] Catella, GC; Burlage, D. "Crystal growth and optical properties of $AgGaS_2$ and $AgGaSe_2$." *MRS Bull.*, 1998, 23, 28–36.

[73] Beasley, JD; "Thermal conductivities of some novel nonlinear optical materials." *Appl. Opt.*, 1994, 33, 1000–1003.

[74] Iseler, GW. "Thermal expansion and seeded Bridgman growth of $AgGaSe_2$." *J. Cryst. Growth*, 1977, 41, 146–150.

[75] Harasaki, A; Kato, K. "New data on the nonlinear optical constant, phase-matching, and optical damage of $AgGaS_2$." *Jpn. J. Appl. Phys.*, 1997, 36, 700–703.

[76] Petrov, V; Noack, F; Badikov, V; Shevyrdyaeva, G; Panyutin, V; Chizhikov, V. "Phase-matching a femtosecond difference-frequency generation in the quaternary semiconductor $AgGaGe_5Se_{12}$." *Appl. Opt.*, 2004, *43*, 4590–4597.

[77] Jackson, AG; Ohmer, MC; LeClair, SR. "Relationship of the second order nonlinear optical coefficient to energy gap in inorganic non-centrosymmetric crystals." *Infrared Phys. Technol.*, 1997, 38, 233–244.

[78] http://www.rp-photonics.com/photodarkening.html. See also Ref. [125].

[79] Korczak, P; Staff, CB. "Liquid encapsulated Czochralski growth of silver thiogallate." *J. Cryst. Growth*, 1974, 24–25, 386–389.

[80] Isaenko, L; Yelisseyev, A; Lobanov, S; Krinitsin, P; Petrov, V; Zondy, J.-J. "Ternary chalcogenides $LiBC_2$ (B= In, Ga; C = S, Se, Te) for mid-IR nonlinear optics." *J. Non-Cryst. Sol.*, 2006, 352, 2443–2443.

[81] Isaenko, F; Yelisseyev, A; Lobanov, S; Titov, A; Petrov, V; Zondy, J.-J; Krinitsin, P; Merkulov, A; Vedenyapin, V; Smirnova, J. "Growth and properties of $LiGaX_2$ (X = S, Se, Te) single crystals for nonlinear optical applications in the mid-IR." *Cryst. Res. Technol.*, 2003, 38, 379–387.

[82] Fossier, S; Salaun, S; Mangin, J; Bidault, O; Thenot, I; Zondy, J.-J; Chen, W; Rotermund, F; Petrov, V; Petrov, P; Henningsen, J; Yelisseyev, A; Isaenko, L; Lobanov, S; Balachninaite, O; Slekys, G; Sirutkaitis, V. "Optical, vibrational, thermal, electrical, damage and phase-matching properties of lithium thioindate." *J. Opt. Soc. Am. B*, 2004, 21, 1981–2007.

[83] Isaenko, L; Yelisseyev, A; Zondy, JJ; Knippels, G; Thénot, I; Lobanov, S. "Growth and characterisation of single crystals of ternary chalcogenides for laser applications." *Opto-Electronics Review*, 2011, 9, 135–141.

[84] Mangin, J; Salaun, S; Fossier, S; Strimer, P; Zondy, J.-J; Isaenko, L; Yelisseyev, A. "Optical properties of lithium thioindate." *Proc. SPIE*, 2001, 4268, 49–57.

[85] Knippels, GMH; van der Meer, AFG; MacLeod, AM; Yelisseyev, A; Isaenko, L; Lobanov, S; Thénot, I; Zondy, J.-J. "Mid-infrared (2.75 - 6.0 μm) second-harmonic generation in $LiInS_2$." *Opt. Lett.*, 2001, 26, 617–619. ($\chi^{(2)}$ for $LiInS_2$ reported as 11.16 pm/V)

[86] Wang, S; Liang, Q; Tao, X; Dekorsy, T. "Optical properties and birefringence in $LiInS_2$ in the terahertz frequency range." *Opt. Mater. Express*, 2014, 4, 575–586.

[87] Nikogosyan, DN. *Nonlinear Optical Crystals: A Complete Survey;* Springer: New York, NY, 2005, p. 265.

[88] Boyd, GD; Kasper, HM; MacFee, JH. "Linear and nonlinear optical properties of $LiInSe_2$." *J. Appl. Phys.*, 1973, 44, 2809–2812. ($\chi^{(2)}$ for $LiInS_2$ reported as 6.8 pm/V)

[89] Yelisseyev, A; Isaenko, L; Lovanov, S; Zondy, J.-J; in Advanced Solid State Lasers, Eds. Injeyan, H; Marshall, C., Vol 34 of OSA Trends in Optics and Photonics Series (Optical Society of America: Washington, D. C., 2000, 561–569. ($\chi^{(2)}$ for $LiInS_2$ reported as 15 pm/V)

[90] Chen, W; Cousin, J; Sigrist, MW; Gao, X; Zondy, J.-J; Isaenko, L; Yelisseyev, A; Lobanov, S. "$LiInS_2$ and $LiInSe_2$: New nonlinear crystals for continuous-wave difference-frequency generation in the mid-infrared." Lasers and Electro-Optics Society, 2006. 19th Annual Meeting of the IEEE, Oct., 2006 (Montreal Canada) 88–89. ($\chi^{(2)}$ for $LiInS_2$ reported as 13.8 pm/V)

[91] Isaenko, L; Vasilyeva, I; Yelisseyev, A; Lobanov, S; Malakhov, V; Dovlitova, L; Zondy, J.-J; Kavun, I. "Growth and characterization of $LiInS_2$ single crystals." *J. Cryst. Growth*, 2000, 218, 313–322.

[92] Andreev, YM; Geiko, LG; Geiko, PP; Grechin, SG. "Optical properties of a nonlinear $LiInS_2$ crystal." *Quantum Electron.*, 2001, 31, 647-648. [transl. from *Kvantovaya Elektron.*, 2001, 31, 1075–1078.]

[93] Petrov, V. "Parametric down-conversion devices: the coverage of the mid-infrared spectral range by solid-state laser sources." *Opt. Mater.*, 2012, 34, 536–554.

[94] Tyazhev, A; Vedenyapin, V; Marchev, G; Isaenko, L; Kolker, D; Lobanov, S; Petrov, V; Yelisseyev, A; Starikova, M; Zondy, J.-J. "Singly-resonant optical parametric oscillation based on the wide band-gap mid-IR nonlinear optical crystal $LiGaS_2$." *Opt. Mater.*, 2013, 35, 1612–1615.

[95] Petrov, V; Zondy, J.-J; Bidault, O; Isaenko, L; Vedenyapin, V; Yelisseyev, A; Chen, W; Tyazhev, A; Lobanov, S; Marchev, G; Kolker, D. "Optical, thermal, electrical, damage, and phase-matching properties of lithium selenoindate." *J. Opt. Soc. Am. B*, 2010, 27, 1902–1927.

[96] Ebrahim-Zadeh, M. "Mid-infrared optical parametric oscillators and applications" in *NATO Science for Peace and Security Series B: Physics and Biophysics. Mid-Infrared Coherent Sources and Applications*, Eds. Ebrahim-Zadeh, M; Gorokina, IT. Springer: Dordrecht, The Netherlands, 2008, 347–375.

[97] Marchev, G; Tyazhev, A; Vedenyapin, V; Kolker, D; Yelisseyev, A; Lobanov, S; Isaenko, L; Zondy, J.-J; Petrov, V. "Nd-YAG pumped nanosecond optical parameteric oscillator based on $LiInSe_2$ with tunability extending from 4.7 to 8.7 μm." *Opt. Express*, 2009, 17, 13441–13446.

[98] Isaenko, L; Yelisseyev, A; Lobanov, S; Petrov, V; Rotermund, F; Slekys, G; Zondy, J.-J. "$LiInSe_2$: A biaxial ternary chalcogenide crystal for nonlinear optical applications in the midinfrared." *J. Appl. Phys.*, 2002, 91, 9475–9480.

[99] Andreev, YM; Badikov, VV; Geiko, PP; Grechin, SG; Efimenko, VV; Efimenko, TD; Panyutin, VI; Tikhomirov, AA; Shaiduko, AV. "Second harmonic generation of TEA CO_2 laser radiation in $LiInSe_2$." *Atoms. Oceanic Opt.*, 2004, 17, 115–120.

[100] Tyazhev, A; Marchev, G; Vedenyapin, V; Kolker, D; Yelisseyev, A; Lobanov, S; Isaenko, L; Zondy, J.-J; Petrov, V. "$LiInSe_2$ nanosecond

optical parametric oscillator tunable from 4.7 to 8.7 μm." *Proc. SPIE*, 2010, 7582, 7580E-1.

[101] Tupitsyn, E; Bhattacharya, P; Rowe, E; Matei, L; Cui, Y; Buliga, V; Groza, M; Wiggins, B; Burger, A; Stowe, A. "Lithium containing chalcogenide single crystals for neutron detection." *J. Cryst. Growth*, 2014, *393*, 23–27.

[102] Schuneman, PG. "Mid-IR nonlinear optical materials" in *Encyclopedia of Materials: Science and Technology* (Second edition), Eds. Buschow, KHJ; Cahn, R; Flemings, M; Ilschner, B; Kramer, E; Mahajan, S; Veyssiere, P. Pergamon: 2001, 5675–5679.

[103] Ni, Y; Wu, H; Huang, C; Mao, M; Wang, Z; Cheng, X. "Growth and quality of gallium selenide (GaSe) crystals." *J. Cryst. Growth*, 2013, 381, 10–14.

[104] Singh, NB; Suhre, DR; Balakrishna, V; Marable, M; Meyer, R."Far-infrared conversion materials: Gallium selenide for far-infrared conversion applications." *Prog. Cryst. Growth Ch.*, 1998, 37, 47–102.

[105] Halasyamani, PS. "Asymmetric cation coordination in oxide materials: Influence of lone-pair cations on the intra-octahedral distortion in d^0 transition metals" *Chem. Mater.*, 2004, 16, 3586–3592.

[106] Kunz, M; Brown, ID. "Out-of-center distortions around octahedrally coordinated d^0 transition metals." *J. Solid State Chem.*, 1995, 115, 395–406.

[107] Ponomarev, VI; Filipenko, OS; Atovmyan, LO; Rannev, NV; Ivanov, SA; Venevtsev, YN. "Structure and lattice dynamics of $SbNbO_4$ crystals at 300-1000 K." *Kristallografiya*, 1981, 26, 341–348.

[108] Kim, YH; Lee, DW; Ok, KM. "Strong second harmonic generation (SHG) originating from combined second-order Jahn-Teller (SOJT) distortive cations in a new noncentrosymmetric tellurite, $InNb(TeO_4)_2$." *Inorg. Chem.*, 2014, 53, 5240–5245.

[109] Lu, H; Gautier, R; Donakowski, MD; Tran, TT; Edwards, BW; Nino, JC; Halasyamani, PS; Liu, Z; Poeppelmeier, KR. "Nonlinear active materials: An illustration of controllable phase matchability." *J. Am. Chem. Soc.*, 2013, 135, 11942–11950.

[110] Kim, SH; Yeon, J; Halasyamani, PS. "Noncentrosymmetric polar oxide material, Pb_3SeO_5: Synthesis, characterization, electronic structure calculations, and structure-property relationships." *Chem. Mater.*, 2009, 21, 5335–5342.

[111] (a) Phanon, D; Gautier-Luneau, I. "Promising material for infrared nonlinear optics: NaI_3O_8 salt containing an octaoxotriiodate(V) anion

formed from condensation of $[IO_3]^-$ ions." *Angew. Chem. Int. Ed.*, 2007, 46, 8488–8491; (b) Ok, KM; Halasyamani, RS. "The lone-pair cation I^{5+} in a hexagonal tungsten oxide-like framework: Synthesis, structure, and second-harmonic generating properties of $Cs_2I_4O_{11}$." *Angew. Chem. Int. Ed.*, 2004, 43, 5489–5491; (c) Li, PX; Hu, CL; Zu, X; Wang, RY; Sun, CF; Mao, JG. "Explorations of new second-order nonlinear optical materials in the K^I-M^{II}-IV-O Systems." *Inorg. Chem.*, 2010, 49, 4599–4605; (d) Sykora, RE; Ok, KM; Halasyamani, PS; Albrecht-Schmitt, TE. "Structural modulation of molybdenyl iodate architectures by alkali metal cations in $AMoO_3(IO_3)$ (A = K, Rb, Cs): A facile route to new polar materials with large SHG responses." *J. Am. Chem. Soc.*, 2002, 124, 1951–1957.

[112] Huang, Y.-Z; Wu, L.-M; Wu, X.-T; Li, L.-H; Chen, L; Zhang, Y.-F. "$Pb_2B_5O_9I$: An iodide borate with strong second harmonic generation." *J. Am. Chem. Soc.*, 2010, 132, 12788–12789.

[113] Ra, H.-S; Ok, KM; Halasyamani, PS. "Combining second-order Jahn-Teller distorted cations to create highly efficient SHG materials: synthesis, characterization, and NLO properties of $BaTeM_2O_9$ (M = Mo^{6+} or W^{6+})" *J. Am. Chem. Soc.*, 2003, 126, 7764–7765.

[114] Phanon, D; Bentria, B; Benbertal, D; Mosset, A; Gauteir-Luneau, I. "New potential materials for infrared nonlinear optics. Preparation, characterisation and optical transparency of monometallic and bimetallic iodates." *Solid State Sci.*, 2006, 8, 1466–1472.

[115] Phanon, D; Mosset, A; Gauteir-Luneau, I. "New materials for infrared non-linear optics. Syntheses, structural characterisations, second harmonic generation and optical transparency of $M(IO_3)_3$ metallic iodates." *J. Mater. Chem.*, 2007, 17, 1123–1130.

[116] Phanon, D; Mosset, A; Gauteir-Luneau, I. "New iodate materials as potential laser matrices. Preparation and characterization of α-$M(IO_3)_2$ (M=Y, Dy) and β-$M(IO_3)_3$ M= Y, Ce, Pr, Nd, Eu, Gd, Tb, Dy, Ho, Er). Structural evolution as a function of the Ln_{3+} cationic radius." *Solid State Sci.*, 2007, 9, 496–505.

[117] (a) Pan, S; Wu, Y; Fu, P; Zhang, G; Li, Z; Du, C; Chen, C. "Growth, structure, and properties of single crystals of $SrBPO_5$" *Chem. Mater.*, 2003, 15, 2218–2221; (b) Pan, SL; Smit, JP; Watkins, B; Marvel, MR; Stern, CL; Poeppelmeier, KR. "Synthesis, crystal structure, and nonlinear optical properties of $Li_6CuB_4O_{10}$: a congruently melting compound with isolated $[CuB_4O_{10}]^{6-}$ units." *J. Am. Chem. Soc.*, 2006, 128, 11631–11634; (c) Zhang, WL; Cheng, WD; Zhang, H; Geng, L;

Lin, CS; He, ZZ. “A strong second-harmonic generation material $Cd_4BiO(BO_3)_3$ originating from 3-Chromophore asymmetric structures.” *J. Am. Chem. Soc.*, 2010, 132, 1508–1509; (d) Huang, YZ; Wu, LM; Wu, XT; Li, JH; Chen, L; Zhang, YF. “$Pb_2B_5O_9I$: An iodide borate with strong second harmonic generation.” *J. Am. Chem. Soc.*, 2010, 132, 12788–12789; (e) Chen, C. “The development of new NLO crystals in the borate series.” *J. Cryst. Growth*, 1990, 99, 790–798.

[118] Huang, Y.-Z; Wu, L.-M; Wu, X.-T; Li, L.-H; Chen, L; Zhang, Y.-F. “$Pb_2B_5O_9I$: An iodide borate with strong second harmonic generation.” *J. Am. Chem. Soc.*, 2010, 132, 12788–12789.

[119] Zou, G; Huang, L; Ye, N; Lin, C; Cheng, W; Huang, H. “$CsPbCO_3F$: A strong second-harmonic generation material derived from enhancement via p-p interaction.” *J. Am. Chem. Soc.*, 2013, 135, 18560–18566.

[120] Ra, H.-S; Ok, KM; Halasyamani, PS. “Combining second-order Jahn-Teller distorted cations to create highly efficient SHG materials: synthesis, characterization, and NLO properties of $BaTeM_2O_9$ (M = Mo^{6+} or W^{6+}).” *J. Am. Chem. Soc.*, 2003, 126, 7764–7765.

[121] Sivakumar, T; Chang, HY; Baek, J; Halasyamani, PS. “Two new noncentrosymmetric polar oxides: synthesis, characterization, second-harmonic generating, and pyroelectric measurements on $TlSeVO_5$ and $TlTeVO_5$.” *Chem. Mater.*, 2007, 19, 4710–4715.

[122] Sasaki, T; Mori, Y; Yoshimura, M; Yap, YK; Kamimura, T. “Recent development of nonlinear optical borate crystals: key materials for generation of visible and UV light “ *Mater. Sci. Eng.*, 2000, 30, 1–54.

[123] Chen, C; Lin, Z; Wang, Z. “The development of new borate-based UV nonlinear optical crystals.” *Appl. Phys. B*, 2005, 80, 1–25.

[124] Ellingson, RJ; Tang, CL. “High-power, high-repetition-rate femtosecond pulses tunable in the visible,” *Opt. Lett.*, 1993, 18, 438–440

[125] Kim, Y; Seo, I; Martin, SW; Baek, J; Halasyamani, PS; Arumugam, N; Steinfink, H. “Characterization of new infrared nonlinear optical material with high laser damage threshold, $Li_2Ga_2GeS_6$.” *Chem. Mater.*, 2008, 20, 6048–6052.

[126] Lin, H; Chen, L; Zhou, L.-J; Wu, L.-M. “Functionalization based on the substitutional flexibility: Strong middle IR nonlinear optical selenides $AX^{II}{}_4X^{III}{}_5Se_{12}$.” *J. Am. Chem. Soc.*, 2013, 135, 12914–12921.

[127] Badikov, VV; Laptev, VB; Panyutin, VL; Ryabov, EA; Shevyrdyaeva, GS. “Study of nonlinear optical characteristics of $AgGa_{1-x}InSe_2$ crystals.” *Quantum Electron.*, 2005, 35, 263–267.

[128] Luo, Z.-Z; Lin, CS; Cui, H.-H; Zhang, W.-L; Zhang, H; He, Z.-Z; Cheng, W.-D. "SHG materials $SnGa_4Q_7$ (Q = S, Se) appearing with large conversion efficiencies, high damage thresholds, and wide transparencies in the mid-infrared region." *Chem. Mater.*, 2014, 26, 2743–2749.

[129] Bera, TK; Jang, JI; Song, J.-H; Malliakas, CD; Freeman, AJ; Ketterson, JB; Kanatzidis, MG. "Soluble semiconductors $AAsSe_2$ (A = Li, Na) with a direct-band-gap and strong second harmonic generation: A combined experimental and theoretical study" *J. Am. Chem. Soc.*, 2010, 132, 3484–3495.

[130] Chung, I; Kanatzidis, MG. "Metal chalcogenides: A rich source of nonlinear optical materials." *Chem. Mater.*, 2014, 26, 849–869.

[131] Chung, I; Malliakas, CD; Jang, JI; Canlas, CG; Weliky, DP; Kanatzidis, MG. "Helical polymer $^1/_\infty[P_2Se_6{}^{2-}]$: Strong second harmonic generation response and phase-change properties of its K and Rb salts." *J. Am. Chem. Soc.*, 2007, 129, 14996–15006.

[132] Chung, I; Jang, JI; Malliakas, CD; Ketterson, JB; Kanatzidis, MG. "Strongly nonlinear optical glass fibers from noncentrosymmetric phase-change chalcogenide materials." *J. Am. Chem. Soc.*, 2010, 132, 384–389.

[133] Bera, TK; Song, JH; Freeman, AJ; Jang, JI; Ketterson, JB; Kanatzidis, MG. "Soluble direct-band-gap semiconductors $LiAsS_2$ and $NaAsS_2$: Large electronic structure effects from weak As···S interactions and strong nonlinear optical response." *Angew. Chem. Int. Ed.*, 2008, 120, 7946–7950.

[134] Bera, TK; Jang, JI; Ketterson, JB; Kanatzidis, MG. "Strong second harmonic generation from the tantalum thioarsenates $A_3Ta_2AsS_{11}$ (A = K and Rb)." *J. Am. Chem. Soc.*, 2009, 131, 75–77.

[135] Gandrud, WB; McFee, JH; Wehmeier, FH. "Nonlinear optical properties of Ag_3SbS_3" *Appl. Phys. Lett.*, 1970, 16, 59–61.

[136] Zhang, Q; Chung, I; Jang, JI; Ketterson, JB; Kanatzidis, MG. "Chalcogenide chemistry in ionic liquids: nonlinear optical wave-mixing properties of the double-cubane compound $[Sb_7S_8Br_2](AlCl_4)_3$." *J. Am. Chem. Soc.*, 2009, 131, 9896–9897.

[137] Zhao, H.-J; Li, L.-H; Wu, L.-M; Chen, L. "Syntheses, crystal and electronic structures, and physical properties of quaternary semiconductors: $Ln_2Mn_3Sb_4S_{12}$ (Ln = Pr, Nd, Sm, Gd)." *Inorg. Chem.*, 2010, 49, 5811–5817.

[138] Chen, M.-C; Wu, L.-M; Lin, H; Zhou, L.-J; Chen, L. "Disconnection enhances the second harmonic generation response: synthesis and

characterization of $Ba_{23}Ga_8Sb_2S_{38}$" *J. Am. Chem. Soc.*, 2012, 134, 6058–6060.

[139] Pamplin, BR; Ohachi, T; Maeda, S; Negrete, P; Elworthy, TP; Sanderson, R. "Solubility of the group IV chalcogenides in I-III-VI_2 compounds." *Inst. Physics Conf. Ser.*, 1977, 35(c), 35–42.

[140] Badikov, VV; Tyulyupa, AG; Shevyrdyaeva, GS; Sheina, SG. "Solid solutions in the $AgGaS_2$-GeS_2 and $AgGaSe_2$-$GeSe_2$ systems." *Inorg. Mater.*, 1991, *27*, 177–180. [transl. from *Dokl. Akad. Nauk SSSR, Neorg. Mater.*, 1991, 27, 248–250.]

[141] Petrov, V; Badikov, V; Shevyrdyaeva, G; Panyutin, V; Chizhikov, V. "Phase-matching properties and optical parametric amplification in single crystals of $AgGaGeS_4$." *Opt. Mater.*, 2004, 26, 217–222.

[142] Miyata, K; Petrov, V; Kato, K. "Phase-matching properties for $AgGaGeS_4$." *Appl. Opt.*, 2007, 46, 5728–5731.

[143] Andreev, YM; Atuchin, VV; Lanskii, GV; Pervukhina, NV; Popov, VV; Trocenco, NC. "Linear optical properties of $LiIn(S_{1-x}Se_x)_2$ crystals and tuning of phase matching conditions." *Solid State Sci.*, 2005, 7, 1188–1193.

[144] Parthé, E. *Crystal Chemistry of Tetrahedral Structures*; Gordon and Breach Science Publishers, Inc.: New York, NY, 1964.

[145] Pauling, L. "The principles determining the structure of complex ionic crystals" *J. Am. Chem. Soc.* 1929, 51, 1010–1026.

[146] Goryunova, NA. *The Chemistry of Diamond-like Semiconductors*; Massachusetts Institute of Technology: Cambridge, MA, 1965.

[147] Pamplin, BR. "A systematic method of deriving new semiconducting compounds by structural analogy." *J. Phys. Chem. Solids*, 1964, 25, 675–684.

[148] Goodman, CHL. "The prediction of semiconducting properties in inorganic compounds." *J. Phys. Chem. Solids,* 1958, 6, 305–314.

[149] Pamplin, B. "The adamantine family of compounds." *Prog. Cryst. Growth Charact.*, 1981, 3, 179–192.

[150] Levine, BF. "Bond-charge calculation of nonlinear optical susceptibilities for various crystal structures." *Phys. Rev. B*, 1973, 6, 2600–2626.

[151] Szymanski, JT. "The crystal structure of cernyite, Cu_2CdSnS_4, a cadmium analogue of stannite." *Can. Mineral.*, 1978, 16, 146–151.

[152] Schäfer, W; Nitsche, R. "Tetrahedral quaternary chalcogenides of the type Cu_2-II-IV-$S_4(Se_4)$." *Mater. Res. Bull.*, 1974, *9*, 645–654.

[153] Rosmus, KA; Brant, JA; Wisneski, SD; Clark, DJ; Kim, YS; Jang, JI; Brunetta, CD; Zhang, J.-H; Srnec, MN; Aitken, JA. "Optical nonlinearity in Cu_2CdSnS_4 and α/β-Cu_2ZnSiS_4, diamond-like semiconductors with high laser-damage thresholds." *Inorg. Chem.*, 2014, 53, 7809–7811.

[154] Rosmus, KA; Aitken, JA. "Cu_2ZnSiS_4". *Acta Cryst.* E., 2011, *67(4)*, i28.

[155] Rosmus, KA; Brunetta, CD; Srnec, MN; Karuppannan, B; Aitken, JA. "Synchrotron X-ray powder diffraction and electronic band structure of α- and β-Cu_2ZnSiS_4." *Z. Anorg. Allg. Chem.*, 2012, 638, 2578–2584.

[156] Lekse, JW; Moreau, MA; McNerny, KL; Yeon, J; Halasyamani, PS; Aitken, JA. "Second harmonic generation and crystal structure of the diamond-like semiconductors Li_2CdGeS_4 and Li_2CdSnS_4." *Inorg. Chem.*, 2009, 48, 7516–7518.

[157] Brant, JA; Clark, DJ; Kim, YS; Jang, JI; Zhang, J.-H; Aitken, JA. "Li_2CdGeS_4, a diamond-like semiconductor with strong second-order optical nonlinearity in the infrared and exceptional laser damage threshold." *Chem. Mater.*, 2014, 26, 3045–3048.

[158] Jang, JI; Park, S; Clark, DJ; Saouma, FO; Lombardo, D; Harrison, CM; Shim, B. "Impact of two-photon absorption on second harmonic generation in CdTe as probed by wavelength-dependent Z-scan nonlinear spectroscopy." *J. Opt. Soc. Am. B*, 2013, 30, 2292–2295.

[159] Wu, Q; Meng, X; Zhong, C; Chen, X; Qin, J. "$Rb_2CdBr_2I_2$: A new IR nonlinear optical material with a large laser damage threshold." *J. Am. Chem. Soc.*, 2014, 136, 5683–5686.

[160] http://www.uqgoptics.com/materials_commercial_corning_pyrexBorosilicate.aspx.

[161] Witte, T; Kompa, KL; Motzkus, M. "Femtosecond pulse shaping in the mid infrared by difference-frequency mixing." *Appl. Phys. B*, 2003, 76, 467–471.

[162] Chatterjee, U; Kumbhakar, P; Chaudhary, AK; Bhar, GC. "Tunable near-infrared radiation by difference frequency mixing in a lithium tetraborate crystal." *Appl. Phys. B*, 2001, 72, 407–409.

[163] http://www.toplent.com/aggase2.htm.

[164] Song, J.-H; Freeman, AJ; Chung, I; Bera, TK; Kanatzidis, MG. "First-principles prediction of an enhanced optical second-harmonic susceptibility of low-dimensional alkali-metal chalcogenides." *Phys. Rev. B*, 2009, 79, 245203.

[165] For some phase-matched THG, see for example Miyata, K. "Pure third-harmonic generation in noncentrosymmetric media." in *Advances in*

Lasers and Electro Optics; Eds. Costa, N; Cartaxo, A. InTech: Rijeka, Croatia, 2010.

[166] Shen, YR. *Principle of Nonlinear Optics*; John Wiley and Sons: Hoboken, NJ, 1984.

[167] Kubelka, P; Munk, F. "A contribution to the optics of pigments." *Z. Tech. Phys*, 1931, 12, 593–601.

[168] Li, Y; Fan, W; Sun, H; Cheng, X; Li, P; Zhao, X. "Electronic, optical and lattice dynamic properties of the novel diamond-like semiconductors Li_2CdGeS_4 and Li_2CdSnS_4." *J. Phys.: Condens. Matter*, 2011, 23, 225401.

[169] McGowan, RJ. "Attenuated total reflectance vs. transmission infrared spectrometry in the quantitative evaluation of paint vehicles." *Anal. Chem.*, 1963, 35, 1664–1665.

[170] Chung, I; Song, JH; Jang, JI; Freeman, AJ; Kanatzidis, MG. "$Na_2Ge_2Se_5$: A highly nonlinear optical material." *J. Solid State Chem.*, 2010, 195, 161–165.

[171] Barnes, NP; Gettemy, DJ; Hietanen, JR; Iannini, RA. "Parametric amplification in $AgGaSe_2$." *Appl. Opt.*, 1989, 28, 5162–5168.

[172] Vodopyanov, KL. "Parametric generation of tunable infrared radiation in $ZnGeP_2$ and GeSe pumped at 3 μm. *J. Opt. Soc. Am. B*, 1993, 10, 1723–1729.

[173] Badikov, VV; Pivovarov, ON; Skokov, YV; Skrebneva, OV; Trotsenko, NK. "Some optical properties of silver thiogallate single crystals." *Sov. J. Quantum Electron.*, 1975, 5, 597–598. [transl. from Kvant. Elektron, 1975, 2, 1090–1092.]

[174] Jang, JI; Clark, DJ; Brant, JA; Aitken, JA; Kim, YS. "Highly efficient infrared optical nonlinearity of a wide-bandgap chalcogenide Li_2CdGeS_4" *Opt. Lett.*, 2014, 39, 4579–4582.

[175] The 2PA coefficient is available at a longer wavelength: see Ref. [153] and Table 3.

[176] Brandi, HS; de Araujo, CB. "Multiphoton absorption coefficients in solids: a universal curve." *J. Phys. C: Solid State Phys.*, 1983, 16, 5929–5936.

[177] He, J; Qu, Y; Li, H; Mi, J; Ji, W. "Three photon absorption in ZnO and ZnS crystals." Opt. Express, 2005, 13, 9235–9247.

[178] Hübener, H; Luppi, E; Véniard, V. "Ab initio calculation of second harmonic generation in solids." Phys. Status Solidi B, 2010, 247, 1984–1991.

[179] Cabuk, S. "The nonlinear optical susceptibility and electro-optic tensor of ferroelectrics: first-principle study." *Cent. Eur. J. Phys.*, 2012, 10, 239–252.

[180] Miller, RC. "Optical second harmonic generation in piezoelectric crystals." *Appl. Phys. Lett.*, 1964, 5, 17–19.

[181] Van Vechten, JA. "Quantum dielectric theory of electronegativity in covalent systems. I. Electronic dielectric constant." *Phys. Rev. B*, 1969, 182, 891–905.

[182] Van Vechten, JA. "Quantum dielectric theory of electronegativity in covalent systems. II. Ionization potential and interband transition energies." *Phys. Rev. B*, 1969, 187, 1007–1020.

[183] Phillips, JC. "Dielectric definition of electronegativity." *Phys. Rev. Lett.*, 1968, 20, 550–553.

[184] Flytzanis, Chr; Ducuing, J. "Second-order optical susceptibilities of III-V semiconductors." *Phys. Rev.*, 1969, 178, 1218–1228.

[185] Levine, BF. "Bond-charge calculations of nonlinear optical susceptibilities for various crystal structures." *Phys. Rev. B*, 1973, 7, 2600–2626.

[186] Levine, BF. "d-electron effects on bond susceptibilities and ionicities." *Phys. Rev. B*, 1973, 7, 2591–2600.

[187] Moss, TS. "A relationship between the refractive index and the infra-red threshold of sensitivity for photoconductors." *Proc. Phys. Soc. B*, 1950, 63, 167–176.

[188] Moss, TS. "Relations between the refractive index and energy gap of semiconductors." *Phys. Status Solidi B*, 1985, 131, 415–427.

[189] Ravindra, NM; Ganapathy, P; Choi, J. "Energy gap-refractive index relations in semiconductors - An overview." *Infrared Phys. Technol.*, 2007, 50, 21–29.

[190] Lin, Z; Jiang, X; Kang, L; Gong, P; Luo, S; Lee, M.-H. "First principles materials applications and design of nonlinear optical crystals." *J. Phys. D.: Appl. Phys.*, 2014, 47, 253001.

[191] Wu, Q; Huang, Y; Meng, X; Zhong, C; Chen, X; Qin, J. "Exploration of new second-order nonlinear optical materials of the Cs-Hg-Br-I System." *Dalton Trans.*, 2014, 43, 8899–8904.

[192] Chen, W; Poullet, E; Burie, J; Boucher, D; Sigrist, M; Zondy, J.-J; Isaenko, L; Yelisseyev, A; Lobanov, S. "Widely tunable continuous-wave mid-infrared radiation (5.5-11 μm) by difference-frequency generation in $LiInS_2$ crystal" *Appl. Opt.*, 2005, 44, 4123–4129.

[193] Jaffe, JE; Zunger, A. "Electronic structure of the ternary chalcopyrite semiconductors $CuAlS_2$, $CuGaS_2$, $CuInS_2$, $CuAlSe_2$, $CuGaSe_2$ and $CuInSe_2$." *Phys Rev. B.*, 1983, 28, 5822–5846.
[194] Zunger, A; Jaffe, JE. "Structural origin of optical bowing in semiconductor alloys." *Phys. Rev. Lett.*, 1983, *51*, 662– 665.
[195] Ozaki, S; Hori, T. "Optical constants and electronic energy-band structure of $AgGaSe_2$." *Jap. J. Appl. Phys.*, 2014, 53, 071202.
[196] Shay, JL; Tell, B; Kasper, HM; Schiavone, LM. "*p-d* Hybridization of the valence bands of I-III-VI_2 compounds." *Phys. Rev B*, 1972, 5, 5003–5005.
[197] Laksari, S; Chahed, A; Abbouni, N; Benhelal, O; Abbar, B. "First-principles calculations of the structural, electronic and optical properties of $CuGaS_2$ and $AgGaS_2$." *Comput. Mater. Sci.*, 2006, 38, 223–230.
[198] Chahed, A; Benhelal, O; Laksari, S; Abbar, B; Bouhafs, B; Amrane, N. "First-principles calculations of the structural, electronics and optical properties of $AgGaS_2$ and $AgGaSe_2$." *Physica B*, 2005, 367, 142–151.
[199] Brunetta, CD; Karuppannan, B; Rosmus, K; Aitken, JA. "The crystal and electronic band structure of the diamond-like semiconductor Ag_2ZnSiS_4." *J. Alloy Compd.*, 2012, 516, 65–72.
[200] Reshak, AH. "Linear nonlinear optical properties and birefringence of $AgGaX_2$ (X=S, Se Te) compounds." *Physica B*, 2005, 369, 243–253.
[201] Hai-Jun, H; Shi-Fu, Z; Bei-Jun, Z; You, Y; Lin-Hua, X. "First-principles calculations of the elastic, electronic and optical properties of $AgGaS_2$." *Phys. Scr.*, 2010, 82, 055601.
[202] Brik, MG. "First-principles study of the electronic and optical properties of $CuXS_2$ (X= Al, Ga, In) and $AgGaS_2$ ternary compounds." *J. Phys. Condens. Matter*, 2009, 21, 485502.
[203] Jaffe, JE; Zunger, A. "Electronic structure of the ternary pnictide semiconductors $ZnSiP_2$, $ZnGeP_2$, $ZnSnP_2$, $ZnSiAs_2$ and $MgSiP_2$." *Phys.Rev. B.*, 1984, 30, 741–756.
[204] Bai, L; Lin, ZS; Wang, ZZ; Chen, CT. "Mechanism of linear and nonlinear optical effects of chalcopyrites $LiGaX_2$ (X = S, Se and Te) crystals." *J. Appl. Phys.*, 2008, 103, 083111.
[205] Ma, T; Sun, L; Xu, C; Chen, Y. "First-principles study of lattice dynamics and thermodynamic properties of $LiInS_2$ (X=S, Se, Te)." *J. Alloys Compd.*, 2011, 509, 9733–9741.
[206] Li, L.-H; Li, J.-Q; Wu, L.-M. "Electronic structures and optical properties of wurtzite type $LiBSe_2$ (B = Al, Ga, In): A first-principles study." *J. Solid State Chem.*, 2008, 181, 2462–2468.

[207] Ma, T.-H; Yang, C.-H; Xie, Y; Sun, L; Lv, W.-Q; Wang, R; Ren, Y.-I. "First-principles calculations of the structural, elastic, electronic and optical properties of orthorhombic $LiGaS_2$ and $LiGaSe_2$." *Physica B*, 2010, 405, 363–368.

[208] Kosobutsky, AV; Basalaev, YM. "First principles study of electronic structure and optical properties of $LiMTe_2$ (M = Al, Ga, In) crystals." *J. Phys. Chem. Solids*, 2010, 71, 854–861.

[209] Yao, JY; Mei, DJ; Bai, L; Lin, ZS; Yin, WF; Fu, PZ; Wu, YC. "$BaGa_4Se_7$: A New Congruent-Melting Nonlinear Optical Material." *Inorg. Chem.*, 2010, 49, 9212–9216.

[210] Yao, J; Yin, W; Feng, K; Li, X; Mei, D; Lu, Q; Ni, Y; Zhang, Z; Hu, Z; Wu, Y. "Growth and characterization of $BaGa_4Se_7$ crystal." *J. Cryst. Growth*, 2012, 346, 1–4.

[211] Chen, S; Walsh, A; Luo, Y; Uang, J.-H; Gong, XG; Wie, S.-H. "Wurtzite-derived polytypes of kesterite and stannite quaternary chalcogenide semiconductors." *Phys. Rev. B*, 2010, 82, 195203.

[212] Xue, D; Zhang, S. "Bond-charge calculation of nonlinear optical susceptibilities of $LiXO_3$ type complex crystals." *Chem. Phys.*, 1998, 226, 307–318.

[213] Xue, D; Zhang, S. "Chemical bond analysis of the correlation between crystal structure and nonlinear optical properties of complex crystals." *Physica B: Condens. Matter*, 1999, 262, 78–83.

[214] Hobgood, HM; Henningsen, T; Thomas, RN; Hopkins, RH; Ohmer, MC; Mitchel, WC; Fischer, DW; Hegde, SM; Hopkins, FK. "$ZnGeP_2$ grown by the liquid encapsulated Czochralski method." *J. Appl. Phys.*, 1993, 73, 4030–4036.

[215] Schunemann, PG; Pollak, TM. "Ultralow gradient HGF-Grown $ZnGeP_2$ and $CdGeAs_2$ and their optical properties." *MRS Bull.*, 1998, 23, 23–27.

In: Nonlinear Optics
Editor: Fred Wilkins

ISBN: 978-1-63321-926-7

Chapter 2

APPLICATIONS OF THE Z-SCAN TECHNIQUE IN ATOMIC MEDIA

Thierry Passerat de Silans*

Laboratório de Superfície - D.F. - CCEN,
Universidade Federal da Paraíba, João Pessoa, PB, Brazil
PPGF, Universidade Federal de Campina Grande,
Campina Grande, PB, Brazil

Abstract

Alkali atoms have large electric dipole moments due to their unique optically active electron in the external shell. Therefore, the coupling between an electric field and the electron is strong, making alkali vapors very suitable media for nonlinear optics studies and applications. Indeed, the observation of nonlinear behavior around atomic resonances is possible with low power of continuous laser light. Moreover, as these resonances are narrow compared to the tuning range of most lasers, one can explore different signs and strengths of the atom-light interaction by only slightly changing the laser frequency. One of the simplest nonlinear phenomenon in alkali atoms originates from a Kerr-like term in the refractive index, an effect involving the lowest nonlinear susceptibility for an center-symmetric media. In the regime where the Kerr effect becomes relevant, this interaction changes the refractive index of the medium, which results

*E-mail address: thierry@otica.ufpb.br

in a modification of the beam propagation in a self-interaction process. For a laser beam with a radial gradient of intensity (as in the typical case of Gaussian beams) a radial gradient of refractive index will appear in such a way that the medium will exhibit a lens-like behavior. This phenomenon, known as self-lensing, is relatively easy to be implemented experimentally, allowing the measurement of the Kerr coefficient of resonant atomic vapors, which appears as an interesting application of the z-scan technique. Indeed, this technique has been extensively used for measurements in liquids and solids, but relatively few experiments have applied it to thermal atomic vapors or cold atoms clouds. In this chapter, we first review the interaction between a laser beam and alkali atoms in order to characterize the origin of the nonlinear refractive index and to explore the rich dependency of such a phenomenon with the frequency detuning around the sharp resonance. Then we discuss the z-scan technique applied to cold atoms clouds and atomic vapors, as well as recent experimental investigations.

1. Introduction

The study of the interaction between light and matter is one of the most important in science. The light-matter interaction can be divided between linear phenomena, for which the optical response of the material is proportional to the electric field amplitude, and nonlinear phenomena, when material optical response varies nonlinearly with the electric field amplitude. For the appearance of nonlinear optical interaction, the light field must be intense in a frequency close to a resonance of the material. Sources of light with high spectral density of intensities has been made available since the demonstration of laser light in 1960 by Maiman [1] and have allowed the emergence of the nonlinear optical field of research. Indeed, the research field of nonlinear optics is usually considered to have began with the second-harmonic-generation experiment by Franken and co-authors [2] in 1961, one year after Maiman's seminal contribution. Nonlinear optics is now a very active research field, resulting in new fundamental research discoveries [3] and technological application to daily life [4].

Most of the research of nonlinear optics has been made using solid and liquid media. Such materials may benefit from the engineering of specially design materials such as, for instance, SBBO crystals for second-harmonic-generation [5]. However, nonlinear interaction of light with solid and liquids often requires high intensity field, usually obtained from pulsed beams. In the last decades, an increase of nonlinear optical research using gaseous media has occurred. In THIS CHAPTER we will concentrate in alkali atoms gases. Below, we discuss some of the advantages of using alkali atomic gases for nonlinear optics experiments:

1. Alkali atoms have a unique electron in their valence band resulting in a large electric dipole moment and a strong coupling to electromagnetic fields. Therefore, nonlinear interaction regime between light and atoms can be attained for low power light. This introduce a practical facility in experimental research relative to nonlinear optical studies in solid and liquid, since one can use low power continuous-waves lasers (cw). For instance, atomic physics has benefits from the development of diode laser devices since the eighties [6]. Nowadays, manufactured diode laser have output power around 100 mW, are tunable over a range of a few tens of GHz, which usually covers the hole structure from ground state of alkali atoms $nS_{1/2}$ to one of the two first fine-structure excited states $nP_{1/2,3/2}$ (transitions usually called D1 and D2 line, respectively). Furthermore, diode laser has the advantages of been compact and presenting low intensity fluctuations.

2. The use of high power cw lasers or pulses in nonlinear experiments with solid and liquid may induce damages on the material. Atomic vapors have the advantages of being non-damageable. After excitation to an excited state, the atom returns to its ground state: i) by radiative or non-radiative decay, collisions for instance, if the atom is not optically pumped; ii) if optical pumping occurs, that is, the atom decays to a ground state different from the one coupled to the laser light, it is re-thermalized by collision with other atoms or with the containing cell surface. The non-damageable character of atomic vapor allows repetitions of the experiments with the same sample. Note that a sample of trapped cold atoms, in opposite, can be damageable by optical forces on the cloud or by modifications of the trapping potential according to the internal atomic state attained [7].

3. Optical properties of solid and liquid samples are usually affected by parameters fluctuations in preparation, impurities, and other defects that decreases precision in experiments repeatability. In opposite, atomic samples are homogeneous media, with characteristics optical nonlinearities

highly repeatable.

4. Atomic resonances are usually sharp. For alkali atoms, for instance, some typical resonances width must be considered: i) the width induced by natural decay of excited levels. Such width is typically a few MHz for the D1 and D2 lines (see values in table 2). ii) for relatively dense atomic vapors and cold atoms clouds, collisions with other atoms increases the linewidth. Typical values of self-pressure broadening for Rb vapor are a fraction of MHz [8] for densities of 5×10^{12} cm^{-3}, that corresponds to typical atomic densities in nonlinear optics experiments. iii) For vapors, the movement of the atoms induces an inhomogeneous broadening of the resonances through Doppler shifts. For vapors at temperatures of a few tens of degrees Celsius to a few hundreds of degrees Celsius typical Doppler width are of hundreds of MHz. iv) Some nonlinear phenomena results in narrower width of the transitions, as it is typically the case for Electromagnetically Induced Transparency (EIT) [9], reaching the regime of some kHz width. All those different width scales are usually well below the tunning range of diode lasers. For different detuning there are different coupling rates between the light beam and the atoms resulting in modifications of the nonlinear phenomena strength. Moreover, detuning the laser frequency to the red or to the blue side of the resonance can change the sign of the nonlinear interaction. Thus, atomic system has the advantage of the possibility to change the nonlinear interaction by a slightly change in the laser frequency, allowing to explore the rich spectral behavior of the interaction. The frequency dependence of the nonlinear interaction can be used for applications as, for instance, the development of a method to stabilize the laser frequency [10, 11] to the atomic resonance. The dependence of the nonlinear interaction with the laser frequency detuning relative to resonance will be widely explored in THIS CHAPTER. Note that liquid and solid resonance are usually thick being difficult to change nonlinear interaction strength just by a change in the laser frequency detuning.

5. Many different level configurations can be explored in the laser interaction with atoms. The most simple configuration is one beam interacting with a two-level atom, giving rise, for instance, to an intensity-dependent refractive index of the vapor or of a cold atomic cloud, main subject of THIS CHAPTER. Many others levels configurations has been explored,

such as three level lambda (Λ) configurations, with two ground states and one excited state [9], double lambda systems [12, 13], five level schemes [14] and so on. Such systems explore different fine, hyperfine and Zeeman sublevels and can be coupled with laser beams of the same wavelength or of different wavelengths.

The many different levels configurations listed above allows exploring a variety of nonlinear phenomena. Typically, in atomic physics, nonlinear phenomena studied are of third-order in electric field since the second-order susceptibility is zero for centrosymmetric systems. Some of the applications of the third-order susceptibility in atomic physics are: i) Third-Harmonic-Generation [15]; ii) EIT [9], iii) four-wave mixing [16]; iV) Self-lensing [17]. We will, in THIS CHAPTER, concentrate in the self-focusing effect that can be obtained from a single beam interacting with the vapor. Self-focusing effect originates from an intensity-dependent refractive index that induces a radial gradient of index of refraction when a beam with spatial intensity structures is incident in the atomic sample. This induced index gradient acts as a lens to the beam itself resulting in changes in the beam spatial profile. Self-lensing is the base of the z-scan technique, one of the most used to measure the third-order susceptibility of a sample [18]. In THIS CHAPTER, we will revise the use of the z-scan technique to measure the intensity-dependent refractive index, and thus the third-order susceptibility, of atomic vapors and of clouds of cold atoms. We will first, in section 2, describe theoretically the interaction of a laser beam with the atomic sample and discuss the origin of the intensity dependent refractive index. In this section we will also explore the rich spectral behavior of self-focusing effect for an atomic sample (see point 4 discussed above). Next, in section 3, we will review the z-scan technique. In section 4, we will review the measurements of the refractive index of atomic alkali samples using the z-scan technique. Finally, we will conclude in section 5.

2. Theory

2.1. The Nonlinear Refractive Index

An electromagnetic wave incident in a media induces a polarization. In nonlinear optics formulation the polarization is usually expanded in power series of the electric field. For a unique beam of angular frequency ω incident on a centrosymmetric media the power series can be written as [19]:

$$P(\omega) = \epsilon_0 \left[\chi^{(1)}(\omega)E^1(\omega) + \chi^{(3)}(\omega)E^3(\omega) + \chi^{(5)}(\omega)E^5(\omega) + \ldots\right], \quad (1)$$

where $E(\omega)$ is the electric field amplitude of a monochromatic beam $\vec{E} = E\hat{e}\left(e^{-i\omega t} + e^{i\omega t}\right)$ ($\hat{e}$ defines the direction of the light polarization), $\vec{P}(\omega) = P(\omega)\hat{e}\left(e^{-i\omega t} + e^{i\omega t}\right)$ is the polarization of the media and ϵ_0 the vacuum permittivity. For a centrosymmetric media, the even orders of the susceptibility are intrinsically equal to zero, thus only the odd orders of the susceptibility appears in equation (1).

The propagation of the light inside the media is governed by wave equation derived from Maxwell's equations:

$$\nabla \times \nabla \times \vec{E} + \mu_0 \frac{\partial^2}{\partial t^2}\vec{D} = 0, \quad (2)$$

with μ_0 the vacuum permeability and $\vec{D}$ is the displacement electric field that is related to the polarization and to the incident electric field by:

$$\vec{D} = \epsilon_0\vec{E} + \vec{P}. \quad (3)$$

Writing the polarization in terms of an effective susceptibility $P = \epsilon_0\chi_{eff}E$ with $\chi_{eff} = \chi^1 + \chi^{(3)}E^2 + \chi^{(5)}E^4 + \cdots$ one obtain:

$$\vec{D} = \epsilon_0\vec{E}\left[1 + \chi_{eff}\right] = \epsilon_0\epsilon(\omega)\vec{E}, \quad (4)$$

with ϵ a relative permittivity that depends on the media optical properties.

The refractive index of the media can be calculated as:

$$n^2 = 1 + \Re\left(\chi_{eff}\right) = 1 + \Re\left(\chi^{(1)} + \chi^{(3)}E^2\right), \quad (5)$$

where it has be written up to second order in the electric field amplitude. $\Re()$ denotes the real part. Equation (5) shows that the refractive index has a term that depends on the square power of the electric field, or in other words, in the intensity of the light beam. The refractive index can thus be written:

$$n = n_0 + n_2 I, \quad (6)$$

where n_0 is the refractive index for low intensity beams, $I = 2n_0\epsilon_0 cE^2$ is the beam intensity, c is the light speed in vacuum and n_2 is usually known as the

Kerr coefficient. By squaring equation (6) and using equation (5) we obtain the Kerr coefficient as a function of the third-order susceptibility:

$$n_2(\omega) = \frac{\Re\left(\chi^{(3)}(\omega)\right)}{4n_0^2\epsilon_0 c}. \tag{7}$$

2.2. How to Calculate the Susceptibility Using Density Matrix Formalism?

To calculate the susceptibility one need to consider the interaction of the electromagnetic wave and the atomic system. Atomic vapors and clouds of cold atoms are ensembles of identical particles in a mixed state. To describe such system it is adequate to use the density matrix formalism. In such formalism, the expectation value of the polarization is given by:

$$\langle P\rangle = Tr\left(\mu\rho\right), \tag{8}$$

where $\langle\rangle$ denotes average over the ensemble, μ is the electric dipole moment of the atomic media, ρ is the density matrix operator and $Tr(A)$ denotes the trace of the operator A. As will be detailed in the sequence we will consider atomic system with one ground state $|0\rangle$ coupled to excited state $|i\rangle$ through dipole interactions, such that the only non-zero elements in the electric dipole moment are $\mu_{0i} = \langle 0|\,\mu\,|i\rangle$ and its complex conjugated. Thus, the polarization is given by:

$$P = N\left(\sum_i \mu_{0i}\rho_{i0} + \sum_i \mu_{i0}\rho_{0i}\right), \tag{9}$$

where N is the atomic density.

From the relation $P(\omega) = \epsilon_0\chi_{eff}(\omega)E$ one can obtain the effective susceptibility from the optical coherence term of the density matrix:

$$\chi_{eff}(\omega) = N\frac{\sum_i \mu_{0i}\rho_{i0} + \sum_i \mu_{i0}\rho_{0i}}{\epsilon_0 E}, \tag{10}$$

from which one can extract the third-order susceptibility by identifying the term that depends quadratically with the electric field amplitude.

In the next subsection we will describe the Hamiltonian of system and calculate the density matrix optical-coherence terms.

Table 1. I, ground hyperfine states and separation for some alkali atoms

Atom	Nuclear spin	Ground hyperfine levels	Ground splitting
^{133}Cs	7/2	F=3,F=4	9.19 GHz
^{87}Rb	3/2	F=1, F=2	6.83 GHz
^{85}Rb	5/2	F=2, F=3	3.04 GHz
^{23}Na	3/2	F=1, F=2	1.77 GHz

2.3. Hamiltonian of the System

Alkali atoms has one electron in the valence band and their ground state is denoted in spectroscopic notations $nS_{J=1/2}$, with n the principal quantum number, S denotes the orbital angular momentum quantum number $l = 0$ and J is the electron total angular momentum quantum number. The fine ground state $J = 1/2$ splits into two hyperfine ground states $F = I \pm 1/2$, where I is the nuclear angular momentum quantum number. The values of I for some alkali atoms are given in table (1).

The separation between ground hyperfine levels is on the order of some GHz while typical cw diode lasers linewidth are of tens of MHz or less. Thus, one can consider that the laser beam only interacts with one of the ground hyperfine levels. The excited $nP_{J=3/2}$ state of alkali atoms has four hyperfine levels: $F' = I - 3/2$; $F' = I - 1/2$; $F' = I + 1/2$ and $F' = I + 3/2$. From one hyperfine ground state, there are three transitions allowed by dipole interaction in the D2 line, since from parity considerations, $F' - F$ should be zero or ± 1, as illustrated in figure (1).

To write the Hamiltonian of the system, we will consider the interaction between a monochromatic electromagnetic field and an atom in the electric dipole approximation. In such approximation, the field amplitude varies very little over the typical dimension of the atom such that the interactions terms involving the field gradient are neglected [20]. The interaction Hamiltonian thus takes the form $H_I = -\mu E$. For parity reasons, the only non-zero elements in the electric dipole moments are those who couples the ground and one excited state (μ_{0i} and μ_{i0}). The non-interacting Hamiltonian can be written $H_0 = \sum_i \hbar\omega_i |i\rangle \langle i|$, with $\hbar\omega_i$ being the excited level energy relative to the ground state. The total Hamiltonian $H = H_0 + H_I$ can be written in the rotating wave approximation

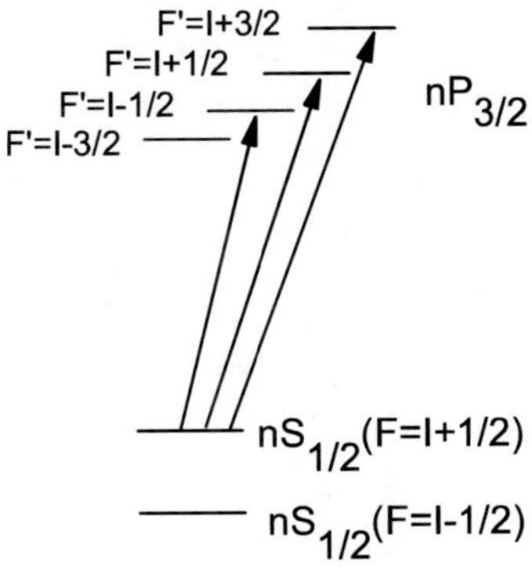

Figure 1. Levels scheme for the D2 transition of an alkali atom. The two ground hyperfine levels are separated by some GHz (see table (1)) and can usually be addressed individually by a cw laser beam. From each ground level three hyperfine excited levels can be coupled by dipole interaction using laser light (shown as arrows in the figure). The levels separations are not shown in scale in the figure.

as:

$$H = \sum_i \hbar\omega_i |i\rangle\langle i| - \sum_i \hbar\Omega_i e^{i\omega t} |0\rangle\langle i| - \sum_i \hbar\Omega_i e^{-i\omega t} |i\rangle\langle 0|, \qquad (11)$$

where $\Omega_i = \frac{\mu_{0i}E}{\hbar}$ is the Rabi frequency of the transition and $\delta_i = \omega - \omega_i$ is the frequency detuning relative to the transition frequency ω_i.

To calculate the optical coherence terms of the density matrix (ρ_{0i}) one has to solve the equation of the evolution of the matrix element:

$$\dot{\rho} = -\frac{i}{\hbar}[H, \rho] + \text{r.t.}, \qquad (12)$$

where the brackets denotes the anticommutation operation and r.t. are relaxation terms not included in the Hamiltonian, such as the decay of excited states due to interaction with the vacuum field.

We will treat in THIS CHAPTER the interaction of laser light resonant with the D2 line of alkali atoms $nS_{1/2} \rightarrow nP_{3/2}$. As mentioned before the D2 line is a four level system with one ground and three excited levels. In the following theoretical development of THIS CHAPTER we will consider the system as a

Table 2. Higher energy ground hyperfine level, separation of the hyperfine excited levels accessible from the ground state by dipole transition (F'=I+3/2, I+1/2 and I-1/2), natural linewidth and Doppler width at room temperature

Atom	Ground state	Excited hyperfine levels separation	Natural width	Doppler width
^{133}Cs	F=4	452 MHz	5.2 MHz	227 MHz
^{87}Rb	F=2	423 MHz	6.0 MHz	307 MHz
^{85}Rb	F=3	184 MHz	6.0 MHz	311 MHz
^{23}Na	F=2	93 MHz	9.8 MHz	790 MHz

closed one such that the excited levels decays only to the ground level coupled by the laser beam, that is, we will ignore optical pumping. The excited hyperfine levels has natural width Γ given by the reciprocal of the mean lifetime of the excited state. For alkali atoms this decay rate is typically a few MHz (see table (2) for such values). Meanwhile the separation between hyperfine excited levels is of tens to hundreds of MHz (see table (2)). In the absence of other broadening effect of the transition, the system can be treated as a two-level atom. This is typically the case for clouds of cold atoms. However, in vapors the atoms are moving with a speed distribution given by the Maxwell-Boltzmann distribution. In this case, each atom sees in his reference rest frame a slightly shifted frequency due to Doppler effect. This results in a broadening of the line known as Doppler or inhomogeneous broadening. Typical values of the Doppler broadening for some alkali atoms at room temperature are given in table (2), being of the order of hundreds of MHz. For Doppler broadened vapors, the laser can interact with the three dipole allowed transition and we have to consider a four-level system with one ground and three excited states. In the following we will calculate the Kerr coefficient for both case: i) two-level systems; and ii) four-level systems.

2.3.1. Two-Level Case

In this section we will calculate the optical coherence matrix elements for a two-level atom and then obtain the expression for the Kerr coefficient. We will first consider atoms at rest (typical for clouds of cold atoms) then we will consider the influence of the movement of the atoms. The analyses of the two-level system will allow to discuss the frequency behavior of the Kerr coefficient. The Hamiltonian for a two-level system is given by:

$$H = \hbar\omega_i \left|i\right\rangle \left\langle i\right| - \hbar\Omega_i e^{i\omega t} \left|0\right\rangle \left\langle i\right| - \hbar\Omega_i e^{-i\omega t} \left|i\right\rangle \left\langle 0\right| \tag{13}$$

which allows to obtain the following optical Bloch equations for the evolution of the density matrix elements:

$$\begin{aligned} \dot{\rho}_{00} &= i\Omega_i\rho_{i0} - i\Omega_i\rho_{0i} + \Gamma\rho_{ii} \\ \dot{\rho}_{ii} &= i\Omega_i\rho_{0i} - i\Omega_i\rho_{i0} - \Gamma\rho_{ii} \\ \dot{\rho}_{0i} &= -i\delta_i\rho_{0i} + i\Omega_i\rho_{ii} - i\Omega_i\rho_{00} - \frac{\Gamma}{2}\rho_{0i} \\ \rho_{00} + \rho_{ii} &= 1, \end{aligned} \tag{14}$$

In the above system of equation Γ is the decay rate of the excited level $|i\rangle$ to the ground state, that is, the homogeneous width of the excited level. The last equation of the system of equations (14) is the conservation of population for a closed system. To write the system of equation (14) we have used the transformation $\rho_{0i} \to \rho_{0i}e^{i\omega t}$. For a cw laser incident in the atomic media, coherence terms reaches the stationary regime and one can solve the Bloch equations imposing $\dot{\rho} = 0$.

Atoms ate rest (cold atoms) Solving the system of Bloch equations (14) and using equation (10) the real part of the effective susceptibility ($\Re(\chi_{eff})$) is obtained:

$$\Re\left(\chi_{eff}\right) = -\frac{4N\mu_{0j}}{E\epsilon_0}\frac{\Omega_j\delta/\Gamma^2}{1+\frac{4\delta^2}{\Gamma^2}+\frac{8\Omega_j^2}{\Gamma^2}}. \tag{15}$$

The term $\frac{8\Omega_j^2}{\Gamma^2} = \frac{I}{I_S}$ is the ratio between the laser beam intensity and the saturation intensity of the atom for excitation at resonance, which is given in table (3) for some alkali atoms. For $\frac{I}{I_S} \ll 1$ one can expand the expression (15):

$$\Re\left(\chi_{eff}\right) = -\frac{4N\mu_{0j}}{E\epsilon_0}\frac{\Omega_j\delta/\Gamma^2}{1+\frac{4\delta^2}{\Gamma^2}}\left[1 - \frac{I/I_S}{1+\frac{4\delta^2}{\Gamma^2}} + \frac{\left(I/I_S\right)^2}{2\left(1+\frac{4\delta^2}{\Gamma^2}\right)^2} - \cdots\right]. \tag{16}$$

In figure (2) we show the contribution of the three first terms in the $\Re\left(\chi_{eff}\right)$ expansion, corresponding to $\Re\left(\chi^{(1)}\right)$, $\Re\left(\chi^{(3)}\right)E^2$ and $\Re\left(\chi^{(5)}\right)E^4$, as a function of the detuning normalized by the homogeneous linewidth, δ/Γ. Also shown in the figure is $\Re(\chi_{eff})$ given by equation (15). The curves are calculated for a ratio $I/I_S = 0.3$, and are shown in linear scale (figure (2)a) and in log-log scale (figure (2)b). One observes the dispersive lineshape of the real part of the susceptibility, the width of such lineshape being narrower the higher the susceptibility order. This is evidenced in the log-log scale where we see that the higher the susceptibility order the faster it falls with the detuning. Actually, for large detuning, $\Re\left(\chi^{(1)}\right)$ falls as δ^{-1}, $\Re\left(\chi^{(3)}\right)$ falls as δ^{-3} and $\Re\left(\chi^{(5)}\right)$ falls as δ^{-5}

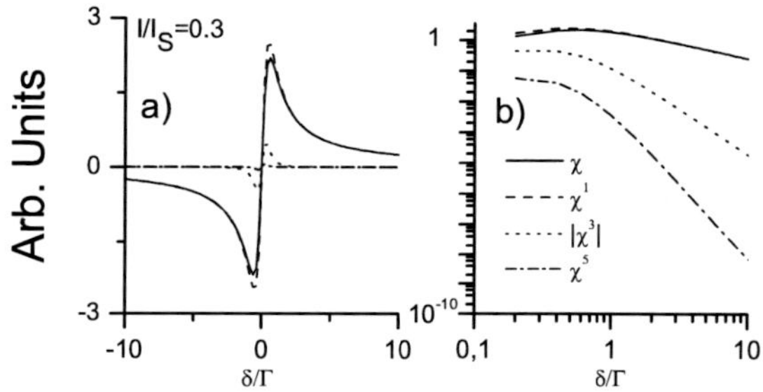

Figure 2. Contribution of the three first terms in the $\Re(\chi_{eff})$ expansion as a function of normalized detuning δ/Γ for a ratio $I/I_S = 0.3$. Also shown is $\Re(\chi_{eff})$ obtained from equation (15). a) Plot in linear scale; b) Plot in a log-log scale to evidence the large detuning behavior.

Alternatively, one can write a detuning dependent intensity saturation $I_S(\delta) = I_S(\delta = 0)\left(1 + \frac{4\delta^2}{\Gamma^2}\right)$ and obtain for the real part of the susceptibility:

$$\Re(\chi_{eff}) = -\frac{4N\mu_{0j}}{E\epsilon_0}\frac{\Omega_j\delta/\Gamma^2}{1+\frac{4\delta^2}{\Gamma^2}}\left[1 - \frac{I}{I_S(\delta)} + \frac{1}{2}\frac{I^2}{I_S^2(\delta)} - \cdots\right]. \quad (17)$$

In figure (3) we show the contribution of $\Re(\chi^{(3)})$ and $\Re(\chi^{(5)})$ from the $\Re(\chi_{eff})$ expansion, the summation of the two first terms of the expansion (equation (16)), as well as, $\Re(\chi_{eff})$ obtained from equation (15), as a function of normalized beam intensity $I/I_S(0)$ ($I/I_S(\delta)$ in brackets). It is clearly seen that truncating the expansion (equation (16)) to the first two terms is only valid for very low beam intensity relative to $I_S(\delta)$. For intensities higher than

$0.25I_S(\delta)$ the third term of expansion (15), corresponding to fifth order susceptibility, is not negligible and the sample does not behaves like a Kerr media. For larger detuning the ratio between $\Re\left(\chi^{(5)}\right)/\Re\left(\chi^{(3)}\right)$ gets smaller and one can consider the media as a Kerr like for higher beam intensities. For instance, for $\delta = 10\Gamma$ the media behaves like a Kerr media for intensities up to $I = 10I_S(0)$ (see figure (3)f).

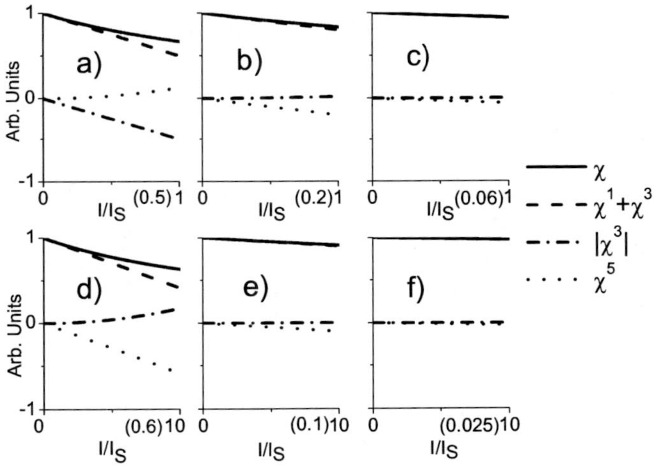

Figure 3. $\Re(\chi^{(3)})$ and $\Re(\chi^{(5)})$ taken from expansion of $\Re(\chi_{eff})$ (equation (16)), the summation of the first two terms of equation (16), as well as, $\Re\left(\chi_{eff}\right)$ calculated from equation (15), as a function of $I/I_S(0)$ (and as a function of $I/I_S(\delta)$ in brackets). a) for $\delta = 0.5\Gamma$; b) for $\delta = \Gamma$; c) for $\delta = 2\Gamma$; d) for $\delta = 2\Gamma$ for beam intensities up to $I = 10I_S$; e) for $\delta = 5\Gamma$ and f) for $\delta = 10\Gamma$.

Table 3. At resonance saturation intensity for some alkali atoms

Atom	I_S
^{133}Cs	1.6 mW/cm^2
^{87}Rb	3.6 mW/cm^2
^{85}Rb	3.9 mW/cm^2
^{23}Na	13.4 mW/cm^2

From the expansion above (16) one obtain the Kerr coefficient to be:

$$n_2(cm^2/W) = 10^4 \times \frac{8N\mu_{0i}^4}{n_0^2\epsilon_0^2 c\hbar^3\Gamma^4} \frac{\delta}{\left(1+\frac{4\delta^2}{\Gamma^2}\right)^2}. \tag{18}$$

Typical lineshape for n_2 is shown in figure (4) as a function of δ/Γ. The Kerr coefficient, n_2, has a dispersive lineshape being negative for red detuning and positive for blue detuning. The maximum and minimum of n_2 occurs at $\delta = \pm\Gamma/\sqrt{12}$.

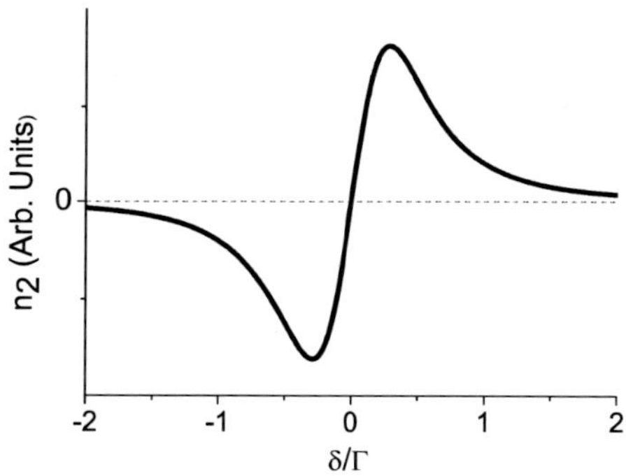

Figure 4. n_2 lineshape for a two level-atom at rest.

Atomic vapor: moving atoms For an atomic vapor, the movement of the atom must be taken into account. An atom with velocity $\vec{v}$ will see the detuning $\delta_v = \delta - \vec{k}\cdot\vec{v}$ in its rest frame. The Kerr coefficient for atom with velocity v_z in the beam propagation direction (that we take as the z axis) is given by equation (18) replacing δ by δ_V. In the vapor, the probability distribution of v_z is given by a Maxwell-Boltzmann distribution, $W(v_z) = \frac{1}{\sqrt{\pi}u}e^{-\frac{v_z^2}{u^2}}$, where $u = \sqrt{\frac{2k_BT}{m}}$ is the velocity distribution half-width, m is the mass of the atom, k_B is the Boltzmann constant and T is the vapor temperature. The Kerr coefficient for an atomic vapor can thus be calculated from:

$$n_2(\delta) = \int_{-\infty}^{\infty} dv_z n_2(\delta_v)W(v_z) \propto \int_{-\infty}^{\infty} \frac{\delta_v}{\left(1+\frac{4\delta_v^2}{\Gamma^2}\right)^2} e^{-v_z^2/u^2}. \tag{19}$$

To obtain the full lineshape of $n_2(\delta)$ one has to calculate the integral numerically. However, analytical solutions can be obtained for small detunings and asymptotically large detuning. For small detuning the main contribution to the integration comes from atoms with velocity $v_z = \frac{\delta}{k} - \frac{\Gamma}{k\sqrt{12}}$, such that $\delta_v = \pm\Gamma/\sqrt{12}$. Those two velocity classes gives contributions with opposite signals to n_2 resulting in a Kerr coefficient proportional to the derivative of the Gaussian Maxwell-Boltzmann distribution [21, 22].

$$n_2 = 10^4 \times \frac{8\pi^{7/2}\mu_{0j}^4 N}{3c\epsilon_0^2 h^3} \frac{\delta}{\Gamma\,(ku)^3} e^{-4\pi\delta^2/(ku)^2}. \tag{20}$$

For large detunings the probability of finding an atom with velocity $v_z = \frac{\delta}{k} - \frac{\Gamma}{k\sqrt{12}}$ is very low due to the exponential decay of the Maxwell-Boltzmann distribution. The main contribution to n_2 comes from atoms with low v_z for which $\delta_v \approx \delta$. That is, the interaction is off resonance and all atomic velocities classes gives the same contribution to n_2 which is given by the large detuning limit of equation: (18).

$$n_2\left(cm^2/W\right) = 10^4 \times \frac{\mu_{0j}^4 N}{2c\epsilon_0^2 \hbar^3 \delta^3}. \tag{21}$$

We show in figure (5) the lineshape of n_2 obtain from numerical integration of equation (19) as well as the curves obtained for low detuning (equation 20) and large detuning (equation 21) asymptotes. This figure clearly shows the rich behavior of the Kerr coefficient with the frequency detuning that can be explored in experiments.

2.3.2. Four-Level Case

The energy level structure of the D2 transition of alkali atoms is not a two level system. Indeed, the excited level $nP_{3/2}$ splits into four hyperfine levels if one consider the coupling between electron total angular momentum and nuclear angular momentum. The hyperfine splitting is larger for heavier atoms. For ^{85}Rb, for instance, the total splitting of the $5P_{3/2}$ level is around 213 MHz. While the Doppler width at room temperature is 310 MHz. As Doppler width is bigger than the hyperfine structure, a two level model is well justified and is consistent with experiments [22, 23], as will be discussed later. This is also true for lighter atoms, for which the hyperfine splitting is lower and the Doppler widths is wider

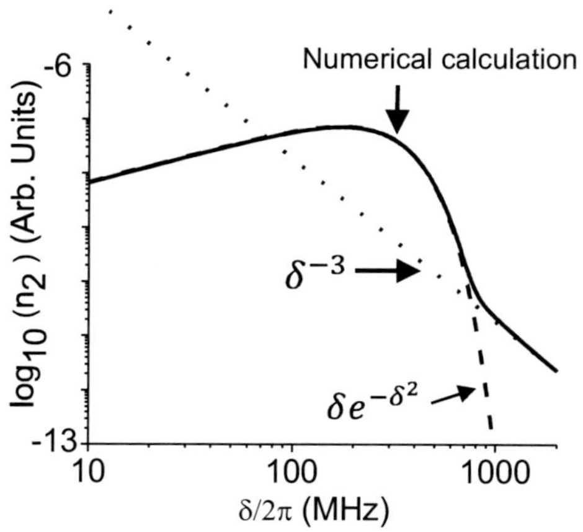

Figure 5. Calculated n_2 for a two level atomic vapor. Solid line shows numerical integration of equation (19), dashed line shows the close to resonance behavior following the derivative of a Gaussian line and dotted line shows the δ^{-3} behavior for far from resonance asymptote.

(the width of the Maxwell-Boltzmann velocity distribution goes as $m^{-1/2}$). Instead, for heavier atoms, the hyperfine splitting becomes on the order of or larger than the Doppler width. For ^{133}Cs, for instance, the total hyperfine splitting is 600 MHz while Doppler width is 227 MHz at room temperature. For such system a two level model is not adequate to treat the light matter interaction close to resonance [21] and one has to consider a closed four-level system consisting of one hyperfine ground state and three hyperfine excited levels accessible by dipole transitions (that is for which $\Delta F = 0, \pm 1$).

The Hamiltonian for the four-level model is given by equation (11). As there is one ground and three excited levels the equation of the evolution of the matrix density gives rise to ten coupled differential equation for density matrix elements, with six of those matrix elements, corresponding to coherence terms, being complex. It is possible to solve the Bloch optical equations analytically, however, the expansion for $I/I_s \ll 1$ is not straightforward and difficult to interpret. A simpler procedure is to make a perturbative expansion of the density matrix $\rho = \rho^{(1)} + \rho^{(2)} + \rho^{(3)}$, with $\rho^{(j)}$ the matrix density of order j in the electric field amplitude [21]. Using such perturbative method the third order susceptibility can be obtained by:

$$\chi^{(3)} = N\frac{\sum_j \left(\mu_{0j}\rho_{j0}^{(3)} + \mu_{j0}\rho_{0j}^{(3)}\right)}{\epsilon_0 E}. \tag{22}$$

Each density matrix order has a time evolution equation given by;

$$\dot{\rho}^{(0)} = -\frac{i}{\hbar}\left[H_0, \rho^{(0)}\right] + \text{r.t.}, \tag{23}$$

$$\dot{\rho}^{(j)} = -\frac{i}{\hbar}\left[H_0, \rho^{(j)}\right] - \frac{i}{\hbar}\left[H_I, \rho^{(j-1)}\right] + \text{r.t., for j} \neq 0. \tag{24}$$

In the zeroth order (without the laser field) the only non-zero density matrix element is $\rho_{00}^{(0)} = 1$. The stationary solution for the first order density matrix for atoms at rest are:

$$\rho_{0i}^{(1)} = \frac{i\Omega_i}{i\delta_i + \frac{\Gamma}{s}} \tag{25}$$

with the others density matrix elements equal to zero.

In second order in the electric field amplitude the population terms and terms of coherence between excited states are non-zero. The population terms are given by

$$\rho_{jj}^{(2)} = \frac{\Omega_j^2}{\delta_j^2 + \Gamma^2/4}. \tag{26}$$

For an atomic vapor one has to integrate the above result over the Maxwell-Boltzmann velocity distribution to obtain the population of each excited state. Close to resonance each excited state population is given by a Gaussian line-shape centered at $\delta_j = 0$.

The coherence between excited levels is given by:

$$\rho_{ij}^{(2)} = \frac{i\Omega_i\Omega_j}{i\left(\omega_j - \omega_i\right)}\left(\frac{1}{\delta_j^2 + \Gamma^2/4} - \frac{1}{\delta_i^2 + \Gamma^2/4}\right). \tag{27}$$

The third order optical coherence term is given by [21]:

$$\rho_{0j}^{(3)} = \frac{-i\delta_j - \Gamma/2}{\delta_j^2 + \Gamma^2/4}\left[2i\Omega_j\rho_{jj}^{(2)} + i\sum_{l\neq j}\Omega_j\rho_{ll}^{(2)} + i\sum_{l\neq j}\Omega_l\rho_{lj}^{(2)}\right]. \tag{28}$$

The first term into brackets of equation (28) is a contribution to the optical coherence between ground level and excited level $|j\rangle$ due to population transfer from the ground state to excited level $|j\rangle$ itself. We will call this first term as the self-population contribution [21]. The second term in equation (28) is due to population transfer from the ground level to the others two excited levels and will be called cross-population contribution. Finally the third term in equation (28) is due to coherence between excited levels.

In figure (6) we plot the self-population, the cross-population and the excited-excited coherence contributions to the Kerr coefficient for a cesium vapor at density $N = 2.4 \times 10^{12}$ atoms/cm^2. Close to resonance the self-contribution dominates while excited-excited coherence contribution is negligible (see figure (6a) in linear scale and figure (6b) in di-log scale). Far from resonance all the three contribution has the same δ^{-3} dependency and are of the same order of magnitude. The main contribution to the self-population comes from the $F = 4 \rightarrow F' = 5$ transition, since it has the higher electric dipole moment (the self-population contribution to $\rho_{0i}^{(3)}$ scales as μ_{0i}^4). The cross-population term has similar contributions from all the three hyperfine transitions. Also show in figure (6a) and (6b) is the four-level model result as a summation of the three contributions of equation (28), as well as, the calculation using the two-level model (equation (19)). As discussed above, the two-level and the four-level models gives different results close to the resonance. Far from resonance the hyperfine structure should not be important and the two models converges to the same values.

3. The Z-Scan Technique

In 1989, Mansoor Sheik-Bahae and co-authors developed a very simple and efficient single beam technique to measure the Kerr coefficient of nonlinear refractive index, named the z-scan technique [18]. In this section, we will review shortly such technique [18, 24] and discuss its application to saturable media. The z-scan technique is based on beam reshaping in the far field after traversing a nonlinear media. A Gaussian beam is focused into the nonlinear sample and is detected after an aperture in the far field. The intensity-dependent refractive index of the media induces a phase profile in the beam at the output of the sample. After propagating, this phase profile induces focusing or defocusing of the beam such that the media act as a thin lens, whose lens-power is related to the Kerr coefficient [25]. The effect of focusing/defocusing can be measured by a change of the intensity transmitted by the aperture in the far-field, thus transforming beam distortion into amplitude variation.

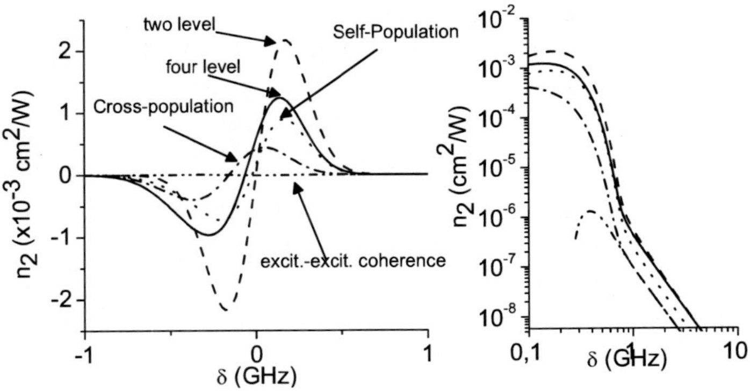

Figure 6. Kerr coefficient for a four-level system as a function of the frequency detuning (solid line) calculated from equation (28) and (22) in a linear scale (a) and in a di-log scale (b). Also shown in figure are: the calculated Kerr coefficient for a two-level model (dashed line); the self-population contribution to the four-level system (dotted line); the cross-population contribution to the four-level system (dashed-dot line); and the contribution to the four-level system from coherence between excited levels. The calculation was made for the D2 line of a cesium vapor at density $N = 2.4 \times 10^{12}$ atoms/cm^2.

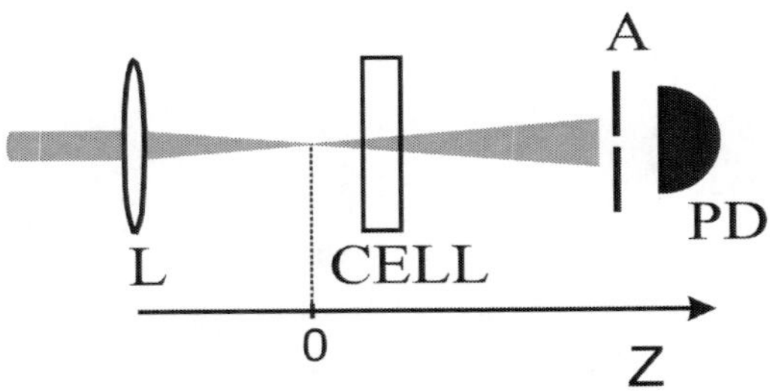

Figure 7. Scheme of the z-scan technique. The beam is focused by a lens (L) and is detected by a photodiode (PD) after an aperture (A). The cell containing the atomic sample is displaced along the z-axis across the focal point.

In the z-scan technique, the nonlinear media sample is placed between the lens and the aperture and is displaced along the beam propagating axis (the z-axis) across the focal point (see figure (7). For positive n_2 the media acts as a convergent lens and a self-focusing effect is observed. When the sample is placed before the lens focal point, the focus will be brought closer to the lens by the self-focusing effect and the beam size will be greater in the aperture plane, reducing its transmission. When the sample is placed after the focal point the self-focusing effect will reduce the beam divergence increasing the aperture transmission. Therefore, the z-scan for a self-focusing media will result in a decrease of the detected signal followed by an increase of the signal, resulting in a dispersive like signal (see figure (8)). For a self-defocusing media, the opposite behavior occurs, that is, one observes an increase of the aperture transmission followed by a decrease of the aperture transmission.

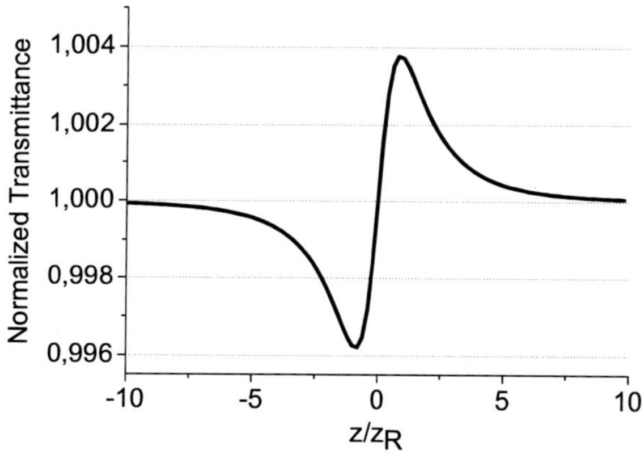

Figure 8. Calculated aperture transmittance as a function of the sample position relative to the beam focal point. The curve was normalized by the transmittance when the sample is far from the lens focal point. The calculation was made for a beam of power 500 μW, focused to a waist of 50 μm and with a Kerr coefficient of $n_2 = 4 \times 10^{-7}$ cm^2/W, which corresponds to the value at $\delta = 1$ GHz in figure (6).

The sign and the magnitude of the Kerr coefficient n_2 of the sample can be obtained from the aperture transmission signal. We will consider a Gaussian beam incident in a thin sample in the sense that there is no distortion of the beam shape inside the nonlinear media. This is achieved when the cell thickness L is much smaller than the Rayleigh range of the beam z_R [24]. The beam at the output window of the cell acquire a phase term with Gaussian structure. Sheik-Bahae and co-authors [24] described the propagation of this phase distorted beam to the aperture by the Gaussian Decomposition Method (GDM) [26] and showed that a simple expression for the transmission through the aperture can be obtained:

$$T = 1 + \frac{4\Delta\phi_0 x}{(x^2+9)(x^2+1)}, \quad (29)$$

with $x = z/z_R$ and $\Delta\phi_0$ is the phase shift on-axis and at the focal point. In the above expression the aperture transmission is normalized to the aperture transmission for the sample far from the focal point, for which there is no self-lensing effect (I/I_s very low). This on-axis at focus phase shift is related to the on-axis at focus beam intensity (I_0), to the sample length, to the beam wavenumber (k) and to the Kerr coefficient [24]:

$$\Delta\phi_0 = kn_2 I_0 L. \quad (30)$$

$\Delta\phi_0$ can be obtained by fitting a theoretical curve to the aperture transmission signal. Moreover, there is simple relationship between $\Delta\phi_0$ and the difference between normalized peak and valley aperture transmitance, ΔT [24]:

$$\Delta T = 0.406\,(1-S)^{0.25}\,\Delta\phi_0, \quad (31)$$

with S the aperture transmittance for the sample far from the focal point.

For z-scan experiments with non-saturable Kerr media it is usual to increase the laser intensity in order to get a greater signal to noise ratio for the aperture transmission signal. However, for saturable Kerr media, as is the case of atomic samples, an increase of the intensity will not increase the signal to noise ratio [27]. Moreover, the radial dependent phase shift induced by the media is no longer Gaussian like but is flattened in the center [27] if one uses intensities close to the saturation intensity. Such change in the phase radial profile modifies the far field beam shape and must be taken into account in the calculation of the aperture transmission. In this case the normalized aperture transmission is given by the expression derived in [28], and the saturation results in a reduction of ΔT. A simplified approach to treat saturable media is to fit the experimental curve by expression (29) and extract a saturated Kerr coefficient $n_2^S(I) = n_2^{nS}\,(1 + I/I_S(\delta))$ [28, 22]. The z-scan experiment is repeated at the same frequency detuning for different intensity values and then the non-saturated Kerr coefficient is extracted from extrapolation of $n_2^S(I)$ for $I \to 0$ [22]. The z-scan equations described above (equations (29) and (31)) are valid for a non-absorbing media. Z-scan experiments with atomic samples are thus usually made for frequency at some Doppler width (or natural width in the case of cloud of cold atoms) from resonance to avoid the strong absorption around the resonance frequency. We will in the next section review the literature of the z-scan experiments for atomic media.

4. Measurement of the Kerr Coefficient of Atomic Sample

4.1. Self-Lensing Experiments with Atomic Samples

Before discussing measurements of the Kerr coefficient of atomic media using the z-scan technique, main subject of THIS CHAPTER, we will breafly review experimental observation of self-lensing effects, as well as others techniques to obtain the Kerr coefficient. The first observation of self-lensing effects with atomic vapor was obtained by Grischkowsky in 1970 [29] using a pulsed beam incident in a potassium vapor. The author has taken photographs of the beam after traveling a 1-m long cell and observed the self-focusing when the central frequency of the pulse was blue detuned relative to the D2 potassium transition. In 1974, Bjorkholm and Ashkin [30], observed self-focusing, self-defocusing and self-trapping of a cw beam in sodium vapor. Using a 20.3-cm-long cell they were able to observe the changes of the beam diameter inside the cell along the beam propagation axis. This was the first demonstration of self-trapping of light due to electronic nonlinear refractive index. R. H. Lehmberg and co-authors has for the first time measured a negative Kerr coefficient n_2 responsible for self-defocusing [31]. They studied two-photon transition between cesium levels $6S$ and $7S$ using 1.06 μm light. Measurement were made by recording the laser beam at a silicon photodiode array after traversing a 1-m-long cell. The observation of this two-photon self-defocusing effect is complicated by others self-defocusing contributions from pulse-induced population changes in the $6S$ and $6P$ levels, as well as, by collision mixing of levels. To be sensitive only to the two-photon electronic contribution the authors used pulsed beams with duration lower than the lifetime of the first excited level and measured $n_2 = -(1.4 \pm 0.2) \times 10^{-30}$N esu for linearly polarized light. For circularly polarized light the main contribution to n_2 comes from induced population changes between $6S$ and $6P$ levels and a value of $n_2 = (0.26 \pm 0.03) \times 10^{-30}$N esu was measured. Besides room temperature and hot atomic vapors, a few investigations were made of transverse effects on the laser beam induced by clouds of cold atoms [32]. Working with cloud of cold atoms has the advantages of: i) being easier to model since one does not have to take into account the effect of the atom movement; ii) having negligible collisions rates and; ii) having a resolved hyperfine structure such that the laser beam only interacts with one ground and one excited level (thus a good two-level system). However, in order to have seizable effects, high densities of the cloud of atoms are needed, being not so common in cold atoms experimental set-up. Furthermore, for a deeper analysis of the nonlinear interaction of a laser beam and the cloud of cold atoms one has to take into account the reshaping of the atomic cloud by the laser [33].

In the examples of experiments cited above the origin of the refractive index is the saturation of the atomic transition, as discussed in section 2. It is worthwhile to mention some experiments where the origin of the nonlinearity comes from optical pumping between ground levels. Suter and Blasberg [34] studied the interaction of a circularly polarized light with an atomic sodium vapor for the generation of transverse solitary waves. Calculation of the beam propagation in the atomic vapor indicates that the initial Gaussian beam can break down due to spatial inhomogeneities of the light when the media has a self-focusing behavior (blue detuned frequencies). However, if diffusion of the atoms are induced by the presence of a buffer gas, such inhomogeneities are smoothed and the initial profile of the beam remains stable. By imaging the beam at the output window onto a photodiode array, the authors has observed self-focusing behavior for blue-detuned light and self-defocusing behavior for red-detuned light. Ackemann and co-authors [25] also studied transverse shaping of a beam traversing a sodium vapor where the nonlinear refractive index is induced by optical pumping. The vapor cell is filled with buffer gas such that the hyperfine structure is not resolved due to pressure broadening. By imaging the output beam into a CCD camera the lens-like behavior of the vapor was studied as a function of beam power and frequency, as well as, of the vapor density. Atomic and radiation diffusion was shown to influence the Kerr coefficient and the output laser beam shape.

4.2. Z-Scan Measurements for Sodium Vapor

The Kerr coefficient of Na vapor was measured using the z-scan technique by S. Sinha and co-authors [35]. The nonlinear refractive index was investigated in the vicinity of both the D1 and the D2 line of Na, corresponding to the wavelengths of 589.6 nm and 589.0 nm, respectively. The authors used a pulsed beam from a dye laser (pulse duration of 30 ns), incident in a cell containing Na at temperatures ranging from 250 °C to 255 °C. Before reaching the cell, the beam passed through a spatial filtering aperture to ensure a Gaussian intensity profile and was focused by a 1-m focal length lens. The beam was attenuated before the cell to ensure that the on-axis phase shift was less than π, to guarantee the applicability of the z-scan technique expression (29). The authors explored frequency detuning ranging from approximately 2.9 to 7.5 Doppler width (Γ_D) and measured values of n_2 of the order of 10^5 esu. The obtained values of n_2 had a frequency dependence in accordance with the δ^{-3} asymptotic limit of a two-level model for detuning larger than 4.5 Γ_D, while for closer to resonance frequencies the measured values deviated from such model. The Doppler width for the D1 and D2 lines at the temperature of the experiment is much larger than the excited hyperfine structure that can be ignored. Moreover, as the laser spectral linewidth is larger than the ground hyperfine splitting and the smallest frequency detuning used is around three times the ground hyperfine splitting, the ground state structure was not taken in account. Disagreement between the used model and experimental results for detuning less than 4.5 Γ_D can possibly be due to not having taken into account the atomic motion effects.

4.3. Z-Scan Measurements for Rubidium Vapor

The non-linear refractive index of Rubidium vapors D2 line was investigated by C. F. McCormick and co-authors [23, 22] using the z-scan technique. The authors used low power spatially filtered continuous wave (cw) laser beam incident in a cell containing natural-abundance rubidium (72% ^{85}Rb and 28% ^{87}Rb) heated at 78 °C (atomic density $\sim 10^{12}$ cm^{-3}). The z-scan measurement explored frequency detuning ranges from 0.8 GHz (2.4 Γ_D) to 1.2 GHz (3.6 Γ_D) at the blue side of ^{85}Rb $5S_{1/2}(F = 3) \rightarrow 6P_{3/2}$ and at the red side of ^{87}Rb $5S_{1/2}(F = 2) \rightarrow 6P_{3/2}$. For this frequency range the measured n_2 was well fitted by a Gaussian like two-level model (see equation. 21) and badly fitted by a two-level δ^{-3} model (see equation 21). The authors considered just one ground hyperfine level, since the ground hyperfine sppliting is much larger than the laser linewidth and the Doppler width. The excited hyperfine structure was ignored since is is much narrower than the Doppler width at the used temperatures (see Table 2 and figure (9)). To ensure a good signal to noise ratio, the authors make the experiment with beam intensities not negligible relative to the saturation intensity and extracted from the fitting of the z-scan curves a saturated Kerr coeficient $n_2^{(S)} = \frac{n_2^{nS}}{1+I/I_S}$ [28]. Repeting the experiment to several laser intensities one can fit the experimental data by the saturation law and extract the non-saturated Kerr coeficient (n_2^{nS}). In the fitting procedure the saturation intensity I_S is kept as a free parameter in order to take into account modifications of the laser-atom interaction due to excited hyperfine structure and due to optical pumping effects between Zeeman sublevels.

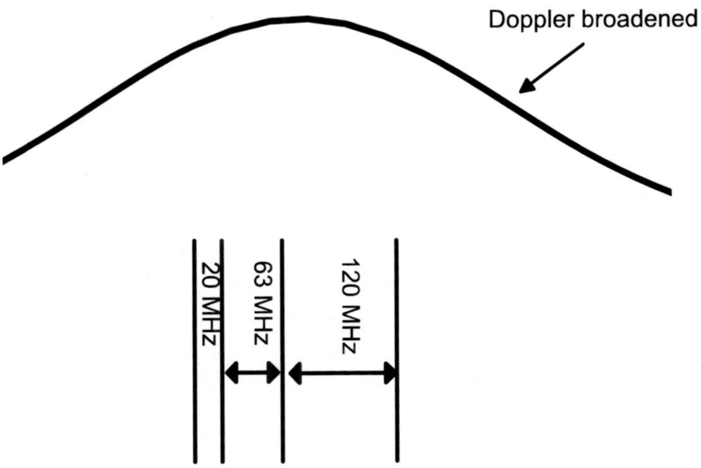

Figure 9. Hyperfine excited levels and the Doppler broadened resonance line at T=300 K for ^{85}Rb.

For laser frequency closer to resonance, linear and non-linear absorption must be considered [23]. Linear absorption modifies the optical power along the z-axis and has influences in the beam propagation inside the vapor. Non-linear absorption depends on the position of the cell along the z-axis since it depends on the intensity of the beam. Close to the focal point, where the intensity is maximum, the negative absorptive nonlinearity, that originates on the absorption saturation, results in higher transmission of the beam through the cell than the transmission far from the focal point. To fit the experimental curves, the authors integrated the wave equation of the beam at each radial point and then propagated the beam until the aperture using a quasi-fast Hankel transform. In the fitting procedure the absorption coefficient at the line center and the saturation intensity are kept as free parameters.

4.4. Measurements for Cold Cesium Clouds

The z-scan technique was also used for studying self-lensing effects of clouds of cold atoms, for instance, clouds of cold cesium were used [33]. One of the advantages related by the authors of using clouds of cold atoms instead of hot vapors is that one needs detuning of a few MHz to obtain detectable self-lensing without absorption effects while with hot vapors one has to detune the frequency for a few GHz. Moreover, with the use of narrow linewidth diode laser, the light-beam interacts only with one excited hyperfine level and the system can be adequately described as a two level atom.

In usual z-scan experiments the atomic sample is moved across the beam focal point. However, the magneto-optical trap (MOT) apparatus used to obtain the cloud of cold atoms is not easy to move. The authors decided to move the focusing lens to realize the z-scan experiment. When moving the lens, the beam focus is displaced across the cloud of cold atoms. For a positive Kerr coefficient, the aperture transmittance is higher before the beam focal point reaches the atomic sample, and is lower when the focal point has passed the cloud of cold atoms. The z-scan signal is modified relative to the usual one by modification of the spatial density profile of the cloud. As the beam focus approach the atomic cloud, radiation forces push the atoms and changes its spatial distribution. Moreover, the atomic sample nonlinearity can change from self-focusing to self-defocusing by the radiation pressure. Indeed, a initially blue-detuned laser relative to resonance for atoms at rest can push the atoms such that, in their reference frame, the laser frequency becomes red-detuned, and thus has a nonlinearity with opposite sign.

4.5. Measurement for Hot Cesium Vapor

M. Araújo and co-authors [21] reported on the measurement of the Kerr nonlinearities for a cesium vapor slightly heated above room temperature. A cw diode laser beam with diameter of 3-mm passes through a 20-cm-focal lens and is detected after passing through an aperture placed at 39 cm after the lens. The transmission through the aperture for non-resonant light is 60%. A cell containing cesium vapor with thickness of 1 mm is displaced across the focal point in order to perform the z-scan measurement. The estimated Rayleight lenght is 5 mm, much larger than the cell thickness to ensure no beam-shape changes occurs inside the cell. The cell is contained inside an oven to be heated at 70 °C in order to increase the atomic density ($N = 2.4 \times 10^{12}$ atoms/cm^3) and improve the signal to noise ratio (as n_2 is proportional to the density).

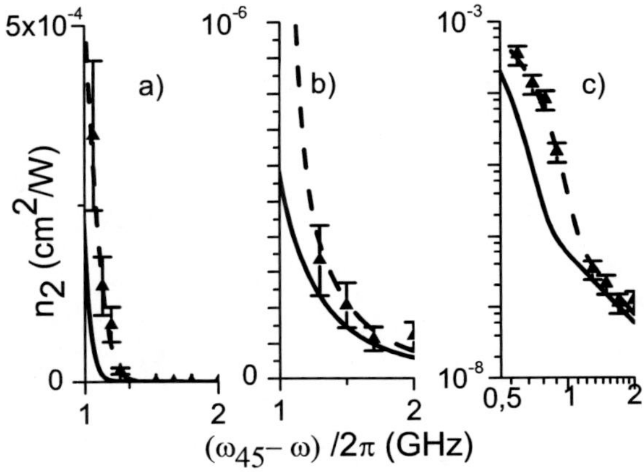

Figure 10. Measured values of the Kerr coefficient at the red side of Cs $6S_{1/2}(F = 4) \rightarrow 6P_{3/2}$ transition. The atomic density of the vapor is $N = 2.4 \times 10^{12}$ atoms/cm^3. Also shown are theoretically calculated values of n_2 using a four-level model (solid line) and a two-level model (dashed line). a) and b) are shown in linear scales while c) is in di-log scale.

The measurements were made at the red side of the closed transition of the D2 line $F = 4 \rightarrow F' = 5$, which has the biggest oscillator strength over the hyperfine structure. The range of detuning measured is from 600 MHz (2.4 Doppler width) to 2000 MHz (8.0 Doppler width), with a change in the Kerr coefficient of almost 4 decades. The figure (10) shows the measured Kerr coefficient in linear (figure 10a and figure 10b) and in di-log scales (figure (10c)) along with theoretical curves obtained from the four-level model (solid line) and the two-level model (dashed line). We clearly see two different qualitative regimes: i) close to resonance, the measured values of n_2 follows the derivative of a Doppler lineshape (see equation (20)), while far from resonance ($\delta > 1200MHz \approx 4.8\Gamma_D$) the measured Kerr coefficient follows a δ^{-3} law given in equation (21) (straight line in a log-log scale). This is the first time to our knowledge that the two different asymptote regimes were observed for the same system. It is also clear from those results that the four-level model is more adequate to describe the nonlinear interaction of a laser beam with Cs atomic vapor than the two-level model, since Doppler width and excited hyperfine levels spacing is of close magnitude. Far from resonance the details of the excited hyperfine structure are not important and four- or two-level models are equivalent.

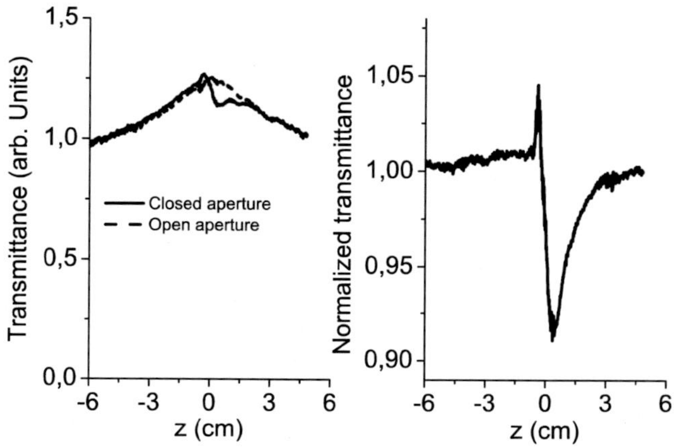

Figure 11. Example of z-scan aperture transmission signal for laser frequency close to resonance. a) Closed aperture (solid line) and open aperture (dashed line) signals. b) Ratio between closed and open aperture signals. The signals were obtained for laser detuning 600 MHz at the red side of Cs $6S_{1/2}(F = 4) \rightarrow 6P_{3/2}(F' = 5)$ transition for a vapor density of $N = 2.4 \times 10^{12}$ atoms/cm^2.

Z-scan measurements close to resonance is not an easy task since linear and nonlinear absorption may play a role. For an open aperture detection configuration, nonlinear absorption will result in a peak-like signal with maximum transmission in the focal point [24, 27]. For a closed aperture configuration, nonlinear absorption and refraction effects are mixed and usually one gets an asymmetric dispersive-like signal [24] or even a peak-like signal with a dispersive lineshape superposed to the peak [23]. For the frequencies close to resonance probed in [21] the absorption peak-like behavior dominates the non-linear refraction effect. To eliminated the nonlinear absorption effect, one can detect simultaneously to the closed aperture transmission an open aperture transmission (see Fig. (11a)) and then normalize the first by the second one (see Fig.

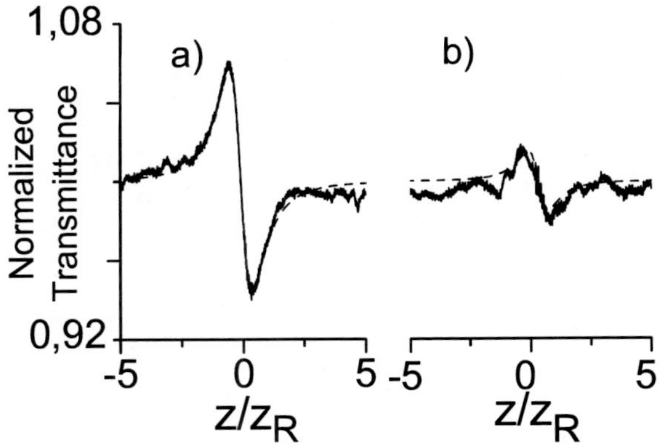

Figure 12. Z-scan signal for frequencies far from the Cs D2 line resonance. a) for $\delta = 1$ GHz and b) $\delta = 1.5$ GHz to the red of the closed hyperfine transition. The solid line is the z-scan experimental signal and the dashed line is a fit using equation (29). The measurement were made for a beam power of 2 mW focused and a vapor density of $N = 2.4 \times 10^{12}$ atoms/cm^2. Solid.

(11b)). This is not a rigorous method but gives reasonable results. In figure (12) we show two z-scan signals obtained for $\delta = 1$ GHz and $\delta = 1.5$ GHz to the red side of $F = 4 \rightarrow F' = 5$ transition. As linear and nonlinear absorption is low for those frequencies the curves exhibits nice dispersive shape. The curves are fitted by equation (29) to extract values of $n_2^{(S)}$. To have a Kerr like media with only third order contribution to the nonlinear polarization, the beam intensity must be much less than the detuning dependent saturation intensity (see figure (3) and equation (17)). However, as the non-linear contribution to the refractive index is $n_2 I$, for low intensity, the signal to noise ratio is not very good. Note that increasing the vapor density to increase n_2 is not a good solution due to the linear and nonlinear absorption. In order to get a signal to noise ratio good enough to extract information, the usual procedure is to repeat the measurements for different beam intensities, fit a value of $n_2^{(S)}$ for each intensity, and then extrapolate the non-saturated Kerr coefficient by fitting a law $n_2^{(S)} = n_2^{(nS)}/(1 + I/I_S)$ (see figure (13)).

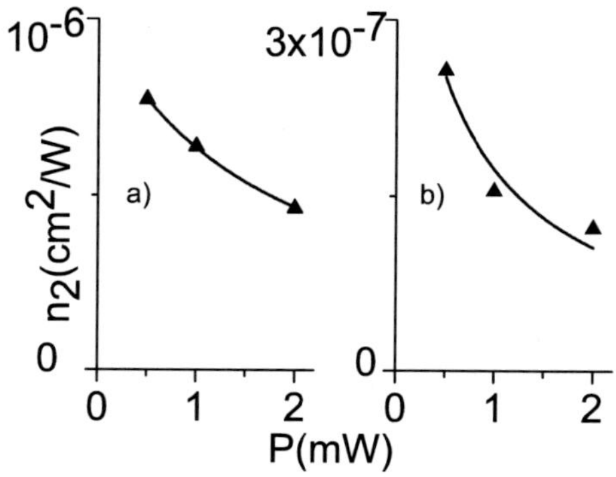

Figure 13. Measured n_2^S value as a function of incident laser power for detuning of a) 1000 MHz and b) 1500 MHz at the red side of transition $6S_{1/2}(F=4) \rightarrow 6P_{3/2}(F'=5)$. The solid lines are fits of equation $n_2^S = n_2^{nS}/(1+I/I_S)$ to the experimental points.

5. Conclusion

Measurement of nonlinear parameters of atomic samples are suitable for many applications. The simplest nonlinear optical effect in atomic media comes from the saturation of the transition induced by a single relatively intense laser beam. This nonlinear interaction results in an intensity dependent refractive index that induces self-lensing effects. The well-established z-scan technique to measure the Kerr coefficient of the refractive index can be used for atomic vapor and clouds of cold atoms. However, linear and nonlinear absorption can be a limitations for the use of z-scan technique for frequencies close to atomic resonances. Moreover, low and moderate laser beam intensities must be used to avoid departure from a pure Kerr like behavior of the interaction. Atomic media has a variety of level structures that allows exploring a rich dependency of the Kerr coefficient with the laser frequency. A few atomic species has been explored up to now (Cs, Rb and Na, to our knowledge), and other sample should be investigated. Besides, nonlinear absorption was poorly investigated in atomic physics and can be an interesting subject of study.

Acknowledgment

The author thanks Brazilian agency CNPQ for financial support.

References

[1] Maiman, T. H. *Nature* 1960, 187, 493–494.

[2] Franken, P. A.; Hill, A. E.; Weinreich, C. W. *Phys. Rev. Lett.* 1961, 7, 118–119.

[3] Kivshar, Y. S. *Optics Express* 2008, 16, 22126–22128.

[4] Garmire, E. *Optics Express* 2013, 21, 30532–30544.

[5] Chen, C.; Wang, Y.; Wu, B.; Wu, K.; Zeng, W.; Yu, L. *Nature* 1995, 373, 322-324.

[6] Wieman, C. E.; Hollberg, L. *Rev. Sci. Instrum.* 1991, 62, 1–20.

[7] Brink, D. M.; Sukumar, C. V. *Phys. Rev. A* 2006, 74, 035401.

[8] Niemax, K.; Movre, M.; Pichler, G. *J. Phys. B: Atom. Molec. Phys.* 1979, 12, 3503–3509.

[9] Boller, K.-J. ; Imamoğlu, A.; Harris, S. E. *Phys. Rev. Lett.* 1991, 66, 2593–2596.

[10] Queiroga,; F. Soares Martins, W.; Mestre, V.; Vidal, I.; Passerat de Silans, T.; Oriá, M.; Chevrollier, M. *Appl. Phys. B* 2012, 313–316.

[11] Martins, W. S.; Cavalcante, H. L. D. de S.; Passerat de Silans, T.; Oriá, M.; Chevrollier, M. *Appl. Opt.* 2012, 51, 5080–5084.

[12] Lukin, M. D.; Hemmer, P. R.; Scully, M. O. *Adv. At. Mol. Opt. Phys.* 2000, 42, 347–386.

[13] Passerat de Silans, T.; Gonçalves, C. S. L. ; Felinto, D.; Tabosa, J. W. R. *J. Opt. Soc. Am. B* 2011, 28, 2220–2026.

[14] Andersen, J. A.; Friese, M. E. J.; Truscott, A. G.; Ficek, Z.; Drummond, P. D.; Heckenberg, N. R.; Rubinsztein- Dunlop, H. *Phys. Rev. A* 2001, 63, 23820.

[15] Young, J. F.; Bjorklund, G. C.; Kung, A. H.; Miles, R. B.; Harris, S. E. *Phys. Rev. Lett.* 1971, 27, 1551–1553.

[16] Liao, P. F.; Bloom, D. M.; Economou, N. P. *Appl. Phys. Lett.* 1978, 12, 813–815.

[17] Chiao, R. Y.; Garmire, E.; Townes, C. H. *Phys. Rev. Lett.* 1964, 13, 479–482.

[18] Sheik-bahae, M.; Said, A. A.; Van Stryland, E. W. *Optics Letters* 1989, 14, 955–957.

[19] Boyd, R. W. *Nonlinear Optics* 3rd ed., Academic Press, USA, 2008, VOl. 1, pp 1–74.

[20] Demtröder, W. *Atoms, molecules and photons: an introduction to atomic-, molecular- and quantum-physics*; Springer: Würzburg, 2006, Vol. 1, pp 242–246.

[21] Araújo, M. O.; Cavalcante, H. L. D. de S.; Oriá, M.; Chevrollier, M.; Passerat de Silans, T.; Castro, R.; Moretti, D. *Phys. Rev. A* 2013, 88, 063818.

[22] McCormick, C. F.; Solli, D. R.; Chiao, R. Y.; Hickmann, J. M. *Phys. Rev. A* 2004, 69, 023804.

[23] McCormick, C. F.; Solli, D. R.; Chiao, R. Y.; Hickmann, J. M. *J. Opt. Soc. Am. B* 2003, 20, 2480–2483.

[24] Sheik-bahae, M.; Said, A. A.; Wei, T.-H.; Hagan, D. J.; Van Stryland, E. W. *IEEE J. Quant. Elect.* 1990, 26, 760–769.

[25] Ackemann, T.; Scholz, T.; Vorgerd, Ch.; Nalik, J.; Hoffer, L. M.; Lippi, G. L. *Opt. Comm.* 1998, 147, 411–428.

[26] Weaire, D.; Wherrett, B. S.; Miller, D. A. B.; Smith, S. D. *Opt. Lett.* 1979, 4, 331–333.

[27] Oliveira, L. C.; Catunda, T.; Zilio, S. C. *Jpn. J. Appl. Phys* 1996, 35, 2649–2652.

[28] Bian, S.; Martinelli, M.; Horowicz, R. J. *Opt. Comm.* 1999, 172, 347–353.

[29] Grischkowsky, D. *Phys. Rev. Lett.* 1970, 24, 866–869.

[30] Bjorkholm, J. E.; Ashkin, A. *Phys. Rev. Lett.* 1974, 32, 129–132.

[31] Lehmberg, R. H.; Reintjes, J.; Eckart, R. C. *Phys. Rev. A* 1976, 13, 1095–1103.

[32] Labeyrie, G.; Ackemann, T.; Klappauf, B.; Pesch, M.; Lippi, G. L.; Kaiser, R. *Eur. Phys. J. D* 2003, 22, 473–483.

[33] Wang, Y; Saffman, M. *Phys. Rev. A* 2004, 70, 013801.

[34] Suter, D.; Blasberg, T. *Phys. Rev. A* 1993, 48, 4583–4587.

[35] Sinha, S.; Bhowmick, G. K.; Kundu, S.; Sasikumar, S; Nair, S. K. S.; Pal, T. B.; Ray, A. K.; Dasgupta, K. *Opt. Comm.* 2002, 203, 427–434.

In: Nonlinear Optics
Editor: Fred Wilkins
ISBN: 978-1-63321-926-7

Chapter 3

NONLINEAR IMAGING MICROSCOPY: METHODOLOGICAL SETUP AND APPLICATIONS FOR EPITHELIAL CANCERS DIAGNOSIS

Javier Adur[1,2,3,*], Hernandes F. Carvalho[2], Carlos L. Cesar[2] and Víctor H. Casco[1]

[1]Microscopy Laboratory Applied to Molecular and Cellular Studies. Engineering School. National University of Entre Rios, Oro Verde, Entre Ríos, Argentina
[2]INFABiC - National Institute of Science and Technology on Photonics Applied to Cell Biology, Campinas, São Paulo, Brazil
[3]CITER - Research and Transfer Center of Entre Ríos, Argentina CONICET-UNER Oro Verde, Entre Ríos, Argentina

ABSTRACT

Nowadays, two-photon excitation and other nonlinear optical approaches have been successfully used within the microscopy framework which evolved as an alternative to conventional single-photon confocal microscopy. In general, they are collectively known as nonlinear microscopy, which includes two-photon excited fluorescence (TPEF)

* Corresponding author: jadur@bioingenieria.edu.ar

microscopy, second, and third harmonic generation (SHG, THG) microscopy, and coherent anti-Stokes Raman scattering (CARS) microscopy. TPEF approach provides high-resolution images of molecular autofluorescence signals in living tissues, SHG microscopy has been successfully used to study structural protein arrays, such as collagen, while CARS and THG signal can be used both to reveal the presence of lipid bodies and to highlight the nuclei inside the thickness of a tissue.

In the first section of this chapter will be described the nonlinear contrast mechanism of each technique, and how to combine and build these setup, on a single commercial multimodal confocal system platform. It will be demonstrate that it is easy-to-operate, suitable to perform TPFE, SHG, and THG imaging, using only the standard femtosecond Ti:sapphire laser. Also, the modifications over the scan head and the detector will be explained and besides the physical evidences of the higher harmonics will be submitted. Following, the usefulness of these techniques to detect early changes associated with the progression of epithelial cancer shall be showed. Thereby, this multi-contrast nonlinear optics microscopy setup, emerge as a valuable tool to differentiate between cancerous, benign, and healthy tissue.

INTRODUCTION

Advances in biomedicine and biology sciences are often assisted by new tools in microscopy. In this sense, multi-photon fluorescence (MPF) or nonlinear optical (NLO) microscopy in recent year has become one of the key imaging modes and evolved as an alternative to conventional single-photon confocal microscopy. One of major area in which NLO microscopy has proven its efficacy is in the study of in vivo tumor biology due to their exceptional characteristics for imaging tissues with minimal invasiveness and chemical selectivity [1]. Several NLO microscopy techniques including, two photon-excited fluorescence (TPEF), second harmonic generation (SHG), third harmonic generation (THG), and coherent anti-Stokes Raman scattering (CARS), alone or in combination has turned out valuable for early detection of precancerous changes [2,3].

The term "nonlinear" is related to the nonlinear dependence of the signal generating processes in use, on the excitation laser intensity. In other words, nonlinear signals depend on the probability of finding more than one photon in space and time, which is considerably enhanced at the focus of a pulsed femtosecond laser. On the other hand, there are nonlinear imaging modalities that produce contrast based on the organization and orientation of nanostructures in the sample. For example, with SHG and THG, specific structural features of the optical configuration of the native specimen generate timely signals. In this case, signal is generated by nonlinearity in the physical properties of the specimen (such as discontinuities in its refractive index). Unlike fluorescence, these harmonic signals are generated without time delay at all, and emerge at wavelengths that are exact integer submultiples of the excitation wavelength and traveling in the same direction [1, 4-6].

Nonlinear imaging provides several advantages: high depth penetration due to use as excitation source a near infrared laser, intrinsic tridimensional sectioning and tridimensional resolution due to the spatial confinement of the signal to the laser focus, multiple nonlinear processes and a multitude of endogenous molecular markers, and low phototoxicity allows the investigation of living processes in the native environment without perturbation [7]. Together, these advantages allow analyzing the complex relations between tissue and organ function and its structure in complex diseases such cancer [8]. Thus, TPEF and THG can be used to image a variety of well-documented morphologic and architectural alterations, including increased nuclear size, increased nucleus/cytoplasm ratio, and loss of normal epithelial architecture. Combined TPEF-SHG microscopies have been applied to the alteration studies both in epithelial cells and the supporting stroma [9]. In particular, SHG has been useful to monitor tumor progression and carcinogenesis, providing information about complex interactions between neoplastic cells, extracellular matrix, and epithelial-stromal communication [10]. Moreover, CARS microscopy provides high-resolution images of the vibrational states of molecules such as symmetric and asymmetric CH-stretching vibration [11, 12]. More relevant, the image processing of this set of NLO microscopy approaches could make it possible to reveal the relationship between signals of epithelial cells and the collagen matrix with early epithelial carcinogenesis [8].

The present chapter presents how combine (TPEF, SHG, THG, and CARS) techniques in one setup to improve the accuracy of multimodal nonlinear images in the detection of epithelial cancer. This chapter first gives an introduction to the principle (characteristic and contrast mechanism) of nonlinear optical microscopy (Section 1). A description of how to combine these techniques and build a multimodal nonlinear platform is given in Section 2. The focus of Section 3 is the demonstration of the physical evidences of the higher harmonics signal whereas Section 4 ends showing the usefulness of these techniques to detect early changes associated with the progression of epithelial cancer.

I. Characteristic and Contrast Mechanism of the NLO Signals

In conventional optical imaging, contrast mechanisms consist of interactions such as absorption, reflection, scattering, and fluorescence, and the response recorded is linearly dependent on the intensity of the incident light. Thereby, there is a linear relationship between the electric field strength of the light and the induced object polarization. At moderately low incident intensity, the optical response can be approximated to be the first-order response as:

$$P = \chi^{(1)} \otimes E \tag{1}$$

where $\chi^{(1)}$: is the linear susceptibility, P: is the polarization of a material and E: is the strength of an applied optical field.

In contrast, nonlinear optical effects used as the imaging contrast, occur when a biological tissue interacts with an intense laser beam and exhibits a nonlinear response to the applied field strength. In this situation, the induced polarization vector P of the material subject to the vectorial electric field E can be expressed as:

$$P = \chi^{(1)} \otimes E + \chi^{(2)} \otimes E \otimes E + \chi^{(3)} \otimes E \otimes E \otimes E + \ldots, \tag{2}$$

where $\chi^{(i)}$: is the i th order nonlinear susceptibility tensor and ⊗: represents a combined tensor product and integral over frequencies [13]. The bulk nonlinear optical susceptibilities $\chi^{(2)}$ and $\chi^{(3)}$ are obtained from the corresponding high-order molecular nonlinear optical coefficients (hyperpolarizability) β and γ by using a sum of the molecular coefficients over all molecule sites. Typically materials with conjugated π-electron structures give large optical nonlinearities. The usual linear susceptibility $\chi^{(1)}$ contributes to the single-photon absorption and reflection of the light in tissues. The $\chi^{(3)}$ corresponds to third-order processes such as two-photon absorption, THG, and CARS, while SHG results from $\chi^{(2)}$ [14].

Figure 1 shows schematically a number of nonlinear optical effects produced for the specimen if the energy density at the focal spot of an objective lens is sufficiently large. Principal contrast mechanisms and characteristics of TPEF, SHG, THG and CARS modalities are described below.

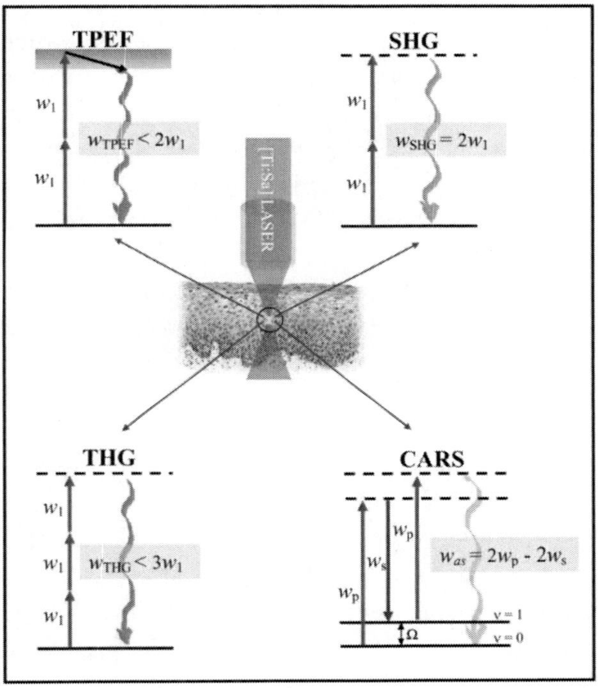

Figure 1. Energy diagrams of NLO techniques. Solid lines represent electronic and vibrational states of molecules, while dashed lines denote virtual states. The straight arrows are excitation beams, whereas wavy arrows, the output signal beams. The black arrow represents relaxation in electronically excited states. ω_1 and ω_2 symbolize the two beams available in CARS microscope. $\omega_p = \omega_1$ and $\omega_s = \omega_2$ in CARS; Ω is a frequency of vibrational transition between vibrational ground state $v = 0$ and vibrationally excited state $v = 1$. ω_1 and ω_2 are pump and Stokes beams from the laser sources, respectively. The ω_3 is a long wavelength beam for THG available on optical parametric oscillator based laser systems for CARS microscope. Abbreviations: SHG: second harmonic generation, THG: third harmonic generation, TPEF: two photon excited fluorescence, CARS: coherent anti-Stokes Raman scattering.

I.1. Two or Multi-Photon Fluorescence Process

TPEF microscopy first appeared by 1990 [15]. It is a third-order non linear optical resonant process where two photons excite an electron from the ground state. It is an inelastic process where photon energy is released at the sample. Two-photon absorption happens only when the energy of the incident photons fall into the two-photon excitation band which is specific for each fluorescent marker. The two-photon excitation band is not exactly half of the one photon excitation band because the selection rules are different. The fact that TPE depends on the square of the incident light provides its confocal characteristics, that is, a process happening only at a focal point volume. Photobleaching is smaller in TPFE compared to single photon excitation because the excited volume is smaller. Therefore, TPEF allow the observation of specific markers, by choosing both, the excitation and emission wavelength [5].

The efficient fluorescence excitation and collection in multi-photon microscopy make it an ideal tool to image deep into turbid biological tissue. Multicolour imaging is allowed to excite different fluorophores simultaneously through different order processes with a single wavelength, in which emissions are spectrally shifted by hundreds of nanometers and uninterrupted for collection. Multi-photon microscopy therefore provides a great flexibility to study tissular physiology and pathology by exciting intrinsic indicators such as autofluorescence; obtaining chemical specificity [14]. Advantages of multi-photon microscopy, in conjunction with animal cancer models, have provided exceptional morphological and functional insights into tumor studies and have opened new approaches to develop novel therapeutics. In particular, the recent advances in semiconductor nanoparticles, such as quantum dots [16] and gold nanorods [17], have opened up another exciting opportunity for multi-photon microscopy to study tumor pathophysiology.

I.2. Harmonic Generation Process

SHG and THG, on the other hand, are coherent second/third-order elastic non linear optical processes. Because two/three photons generate another photon with two/three times the energy of the incident photons, there is no energy released to the medium, meaning no in focus cell photodamages are expected from these processes. Both SHG/THG can be isolated from fluorescence by the wavelength or even by time gating, because the coherent processes are practically instantaneous. The fact that SHG signal is proportional to I^2, while THG signal is proportional to I^3, where I, is the incident light intensity, provides confocality for both techniques. For the same wavelength of the incident light, THG has better optical sectioning resolution than SHG or TPEF, but is also more sensitive to changes in the intensity of the light in the focused spot, such as those caused by laser instability or by scattering or defocusing of the illumination [18]. Doing a bit of history, the first biological image using SHG was obtained by Freund in 1986 [19], after that, in the late 90´s Alfano´s group published several papers showing TPEF, SHG and THG on biological samples [20-23]. To our knowledge, the first laser scanning high resolution SHG image was obtained in 1999 by Campagnola [24]. THG tomography (x vs z) of optical fiber glasses was observed in 1997 [25] and then, Chu and co-workers showed a multimodal imaging including TPEF+SHG+THG in 2001 [26].

Usually the third-order nonlinear susceptibility (χ^3), responsible for THG, is much smaller than the second-order nonlinear susceptibility (χ^2), responsible for SHG. In principle, this would mean that THG should be much harder to observe. However, χ^2, as well as any other even susceptibility coefficients, must be null in the presence of inversion symmetry. Therefore, SHG shall be zero in the presence of centro-symmetric molecules, unless an external parameter, such as electric fields or interfaces, breaks the symmetry. On the other hand, all materials have non-zero third-order susceptibility χ^3. Moreover, χ^3 can be several orders of magnitude larger or smaller for different materials. However, THG is null in a homogeneous material, no matter how high χ^3 could be, because the Gouy-phase shift of π across the focus of a Gaussian excitation beam creates a destructive interference between signals generated before and after the focus [27]. Nevertheless, for a non-homogeneous focal volume a measurable amount of third harmonic is generated [28]. Because heterogeneity is more common in biology than homogeneity, THG provides an important tool for bioimaging, with the warning that it tends to be brighter at the interface of large granules, droplets or other biostructures compared to the internal signal [14].

SHG imaging modality can probe molecular organization, molecular symmetry, orientation, molecular alignments and ultrastructures on the micro- as well as the nanoscale. Because most biological structures are not highly ordered, they are optically isotropic and do not produce any SHG signal. Only those few biological structures that are ordered or that involve some spatial organizations that break the optical centro-symmetry can produce harmonic signal. Namely collagen, the major protein of the extracellular matrix, is one of the best-known SHG structures in biology. Collagen fibrils often aggregate into larger, cable-like bundles, several micrometers in diameter. This regularly staggered packing order provides the needed structural conditions for efficient

SHG [29]. Other examples are acto-myosin complexes in muscle and microtubular-based structures in living cells [30, 31].

Discontinuities in refraction index of the optical dispersion properties of biological tissues are able to generate THG signals [32, 33]. The THG can be used to study optical cell interfaces such as those at cell membranes or organelle surfaces. For example, the surface of the erythrocyte can generate significant THG [34].

In contrast to the chemical specificity that characterizes fluorescence images, harmonic generation (SHG and THG) provides an imaging modality specific for structural configuration. Our experience with both techniques in the study of cancer tumors is that SHG is an excellent tool to observe collagen network of extracellular tissue while THG allow to clearly display the nuclei, which are two important parameters for pathologists [3, 8, 35, 36].

I.3. CARS Process

In addition to harmonic generation microscopy, CARS microscopy is another 3D high resolution imaging approach that circumvents exogenous probes. CARS is a four-wave mixing process in which a pump beam at frequency w_p, a Stokes beam at frequency w_s and a probe beam at frequency w'_p are interacted with a sample to result in an anti-Stokes signal at $w_{as} = 2w_p - w_s$. In nearly all experiments, the pump and the probe beams are derived from the resonant oscillation when the beat frequency (w_p - w_s) matches the frequency of a particular Raman active molecular vibration mode. Furthermore, due to its coherence nature, CARS signal production only occurs when the field-sample interaction length is less than the coherence length. The generated CARS signal is proportional to $(\chi^{(3)})^2\ I^2_p\ I_s$, having a quadratic dependence on the pump field intensity and a linear dependence on the Stokes field intensity. It therefore provides a 3D sectioning capability in CARS microscopy [14].

First CARS microscopy setup was described at the nineties [37] and has now matured into a powerful method for biological imaging. CARS microscopy is more informative than SHG and THG microscopy since it contains rich spectroscopic information about specific molecular species. In cancer applications, CARS was used to classified normal and malignant breast tissue [38], to imaging lipid in cancer metastasis [39], and has proved to be useful in imaging of tissue and identification of cancers in brain, lung and kidney [40, 41], just as few examples.

To summarize this first section, a description of the physic properties, characteristics, and principal contrast mechanisms of each nonlinear optical imaging method described are summarized in Table 1.

Table 1. Characteristics of nonlinear optical microscopy

Characteristics	TPEF	SHG	THG	CARS
Discovery in optics	1961	1961	1962	1965
Application in bioscience	1990	1986	1997	1982
Number of photons	2	2	3	3
Susceptibility	χ^3	χ^2	χ^3	χ^3
Advantages	Deeper imaging with less phototoxicity. Spatial localization for Fluorescence excitation.	Coherent process, symmetry selection. Probing well-ordered structures, functions of membranes, nonfluorescence tissues. No absorption of light.	Coherent process, no symmetry requirement. No absorption of light. Imaging both in bulk and at surfaces for extended conjugation of pi electrons.	Coherent process. Inherent vibrational contrast for the cellular species, requires no endogenous or exogenous fluorophores. Vibrational and chemical sensitivities.
Contrast mechanics	Electronics levels of the molecules	Nonlinear properties of the medium	Nonlinear properties of the medium	Vibrational levels of the molecules
Information	Autofluorescenc e of some biological substances (NADH, FAD, etc.)	Non centrosymmetri c molecules with spatial organization (Collagen, elastin, etc.)	Interfaces, optical inhomogeneities, (Cell edges, lipids, membranes)	Chemical information (lipids, DNA, proteins)

II. Multimodal Nonlinear Optical Microscopy Setup

In recent years there have been significant advances in new technologies used in nonlinear microscopy, i.e., laser sources, scanning systems and detectors. In this context, today there are multiples ways to assemble a nonlinear microscopy platform. The engineering challenge is to integrate the different modalities on a single platform. In response to this challenge, manufacturers have designed microscopes with multiple input and output ports and increased infinity space for the introduction of customized optics. Here, first we enumerate some laser sources, detectors and confocal body microscopy habitually used in this technology and describe in detail two possible setup configuration; one build around an old scan head confocal model and other build in one of the latest scan head models launched in the market.

For high harmonic generation and multiphoton fluorescence microscopies, short fs-pulses of high peak power are required. While ultrashort few cycle pulses are spectrally very broad, they allow for simultaneous excitation of different chromospheres with spectrally separated absorption bands. Available laser with this characteristic are: the titanium:sapphire (Ti:Sa) (wavelength range 700-980 nm, pulse width 100 fs, and 76-100 MHz repetition rate), the Cr:forsterite laser (wavelength range 1230-1270 nm, pulse width 65 fs, and 76-120 MHz repetition rate), the Nd:glass laser (wavelength range 1053-1064 nm, pulse width 150 fs, and 70-150 MHz repetition rate), and the femtosecond ytterbium laser (wavelength range 1030 nm, pulse width 200 fs, and 50 MHz repetition rate). The SHG wavelength excited by a Ti:Sa femtosecond laser operating at 940 nm will be in the blue at 470 nm, while the TPEF will be in the visible region (> 470nm). While that THG signal generated by a 940 nm principal beam, will fall at 313 nm, in the UV region. As a result, the THG signal will suffer from the high UV absorption of the principal biological specimens making signal detection difficult. In contrast, using the Cr:forsterite laser operating in the range of 1230 nm, allows SHG (615 nm), THG (410 nm), and TPEF (> 615 nm), all to fall within the visible spectrum. Additionally, the lowest light attenuation in biological material is generally found in the 1000 to 1300 nm. In recent years appeared other ultrafast laser system, such InSight™ DeepSee™ (wavelength range 680-1300 nm, pulse width <120 fs, and repetition rate 80 MHz), that also can be an excellent light source for multi-modality microscopy. Moving the excitation wavelength to 1200 nm, not only the visible but also the NIR spectrum is open for signal recording.

Nonlinear microscopes share many common features with confocal laser scanning microscopes. In fact, many research groups have implemented multimodal nonlinear platform by coupling source lasers described previously into a confocal scanning microscope. Practically all scan head models of different manufactures have been used with this propose. Just to mention a few, Olympus FV300 [5], Olympus FV1000 [42], Zeiss LSM Meta 510 [43], Zeiss LSM 710 [44], Nikon C1 [45], and Leica TCS-SP5 [46].

In general, the generated nonlinear signals can be collected with the same microscope objective, separated by a dichroic mirror, which is expressly selected for the given fundamental and fluorescence or harmonic emission wavelengths and focused with a lens through the filter onto detector. If Ti:Sa laser is used, the wavelengths are well within in the sensitivity range of high quantum efficiency (QE) silicon-based detectors and PMT photocathodes, that are the detectors normally used. If the source laser used are Nd:glass or Cr:forsterite special NIR detectors (ie: InGaAs photodiode) are need. For more data about the optical characteristics of the different detectors, which is beyond the scope of this chapter, the readers can find excellent information in [18].

Below we describe the two setups built at National Institute of Science and Technology on Photonics Applied to Cell Biology (INFABIC) at the State University of Campinas, on commercials confocal system platforms (inverted Olympus IX 81, FV300 confocal and inverted Zeiss Axio Observer.Z1 and confocal LSM 780).

II.1. Microscopy Setup for TPEF, SHG, and THG

The schematic of the setup is shown in figure 2. The femtosecond laser is a Ti:Sapphire Mai Tai HP Spectra-Physics (Irvine, USA), which provides <100 fs pulses from 690 to 1,040 nm with 80 MHz repetition rate and output power from 1 to 3.5 W. It was equipped with a DeepSee for group velocity dispersion compensation and a broadband halfwave plate (HWP), coupled to a calcite polarizing beam splitter (PBS), as an attenuator. The beam is coupled to the scan head (Olympus FV300) through a custom made port and a dichroic mirror. A collimating telescope (L1, L2) is used to adjust both the beam diameter to fill the objective back-aperture of the high numerical aperture oil immersion objective and the beam focus position on the microscope focal plane. An Olympus FV300 scanner and the laser combiner FV-5 COMB 2 by Olympus (Olympus, Tokyo, Japan) system built onto an Olympus inverted research microscope IX-81 is employed in the setup. The FV300 scanner unit has two input ports, one for a CW laser and the other specifically designed for the pulsed Ti:Sapphire laser. The other components in the excitation pathway are: dichroic mirror to couple both, CW and pulsed lasers, excitation dichroic mirror, XY galvanometer mirror scanners and pupil lens. The excitation for confocal microscopy is provided by an Argon ion laser. Light from the laser source is brought into the scan head using an optical fiber cable. The inverted IX-81 microscope used in the setup is a fully motorized advanced research microscope.

The system offers standard three channel detection, two internal (inside of scan head) and one external, for transmitted light imaging. The epi-detected fluorescence light is collected with the objective, descanned at the galvo-mirrors, separated from the excitation by a dichroic emission beam splitter and focused on the pinhole by a collector lens. Our pinhole turret has 60, 100, 150, 200 and 300 μm confocal apertures. A dichroic filter after the pinhole allows the selection of two wavelength windows, each one with a barrier filter in front of the internal photo multiplier detector to completely block the laser. The microscope back ports and two side ports were used for harmonic detection.

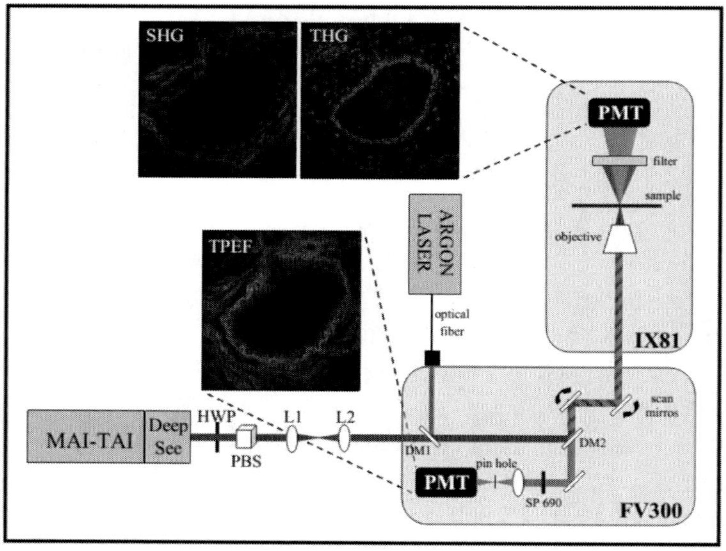

Figure 2. Schematic setup of a multimodal NLO microscope using a femtosecond laser source. The Mai Tai output at 940 nm is couple to the FV300 confocal scanning head and inverted microscope Olympus IX-81 (Olympus, Japan). The 940-nm beam is used for TPEF, SHG, and THG imaging. One internal PMT detector is used for TPEF backward analysis and other PMT external detectors is added for forward SHG and THG detection. Representative TPEF (green), SHG (red) and THG (magenta) breast images obtained with the setup are shown. Abbreviations: HWP: half wave plate, PBS: polarizing beam splitter, L1/L2: telescope lens, DM1 and DM2: dichroic mirrors, SP: short pass filter, PMT: photomultiplier tubes, TPEF: two photon excited fluorescence, SHG: second harmonic generation, THG: third harmonic generation.

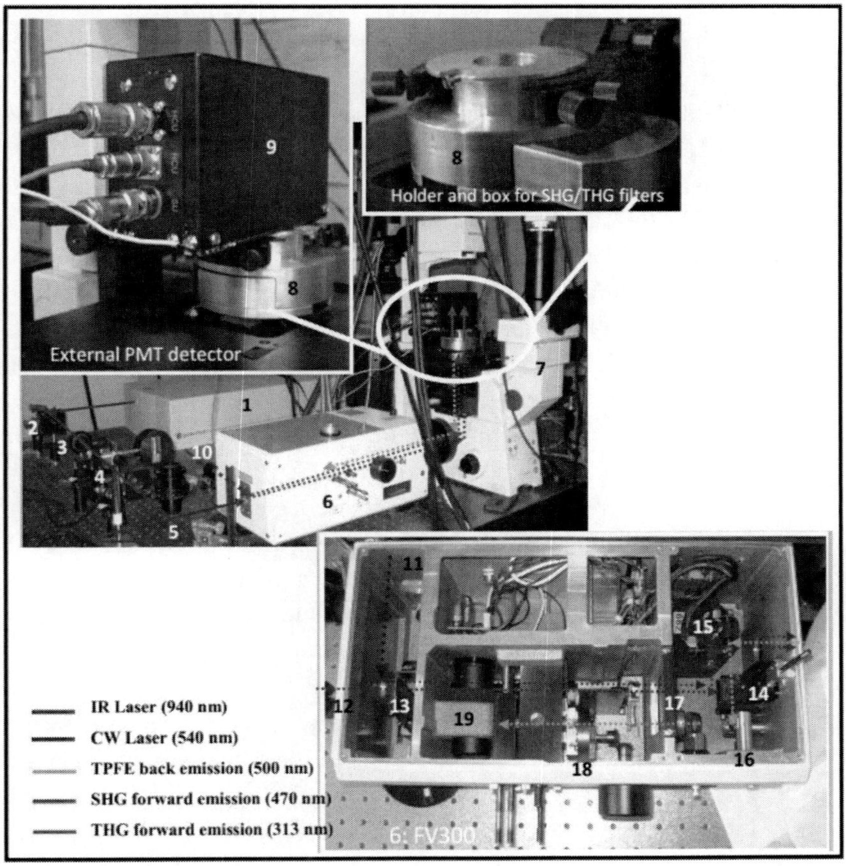

Figure 3. Real setup showing in the center: laser way and Olympus FV300 confocal scanning head coupled to the inverted microscope Olympus IX-8; in the top: external PMT and special holder and in the bottom: internal vision of scan head. Numbers references: 1: Mai-Tai laser and DeepSee module, 2: reflected IR mirror, 3: half wave plate and polarizing beam splitter, 4: telescope, 5: reflected IR mirror, 6: scan head, 7: inverted microscopy, 8: holder and box filters, 9: external PMT, 10: Argon laser, 11: input port to Argon laser, 12: input port to Mai Tai, 13-14: dichroic mirrors, 15: galvanometer mirror scanners, 16: collector lens, 17: short pass filter, 18: pinhole, 19: internal PMT. Abbreviations: IR: infrared, CW: continuous wave, TPEF: two photon excited fluorescence, SHG: second harmonic generation, THG: third harmonic generation.

NLO images were acquired with the 940 nm excitation, generating a TPFE above 500 nm, SHG at 470 nm and THG at 313 nm. TPEF signals were detected in backward configuration, and were detected with the internal scan head PMT (R3896 photomultiplier tubes (PMTs) (Hamamatsu Photonics, Hamamatsu City, Japan)) after a blocking filter (SP) E-690-SP (Omega Filters) to prevent back-reflected laser light with a completely open pinhole. SHG and THG signals were collected in forward direction using an external detector. THG UV signals are generated at the sample but can be absorbed by optical elements in the beam pathway. To avoid this we decided to place the detector right after the sample. Thus, the microscope condenser was replaced by a special home-made PMT support, holding it as close to the sample as possible, without touching it, but with enough space for the optical filters (figure 3, top photographs). The PMT large area assured a good numerical aperture collection angle. To detect SHG signal an E700-SP short pass filter was used, followed by a narrow (10 nm FWHM) bandpass optical filter centered at half the excitation wavelength (475BP, Bio-Rad) to reject the excitation laser beam and any unwanted fluorescence. To collect THG images both SHG filters were replaced with two colored glass filters U340±30 nm (Hoya Corporation). SHG and THG images were sequentially acquired by the exchange of the optical filters, but simultaneously with the TPEF images. Both TPEF were compared to make sure that the sample did not move during the optical filter exchange procedure. Except for the exchange of filters, the system is very friendly to use.

A whole representation of the different light paths, filters, dichroic mirrors and detectors distributions that compose this multimodal system can be found in figure 3. Blue and dark red lines indicate cw Argon laser and pulsed NIR excitation light directed onto the sample for confocal and NLO modality respectively. Green lines indicate fluorescence emission light from confocal and epifluorescence modes. Red and blue lines indicate forward propagating SHG and THG signal respectively. All TPFE, SHG, and THG images showed in figure 2 were acquired with 512 × 512 pixel spatial resolution, using 5ms for the pixel dwell time, with total scanning time of the order of 3s, after a 5 frames Kalman filtering. Images (pseudo-green TPEF, pseudo-red SHG, and pseudo-magenta THG) were combined into a single image for visualization using Olympus Fluoview Image Browser.

II.2. Microscopy Setup for TPEF, SHG, THG, and CARS

The schematic of the setup is shown in figure 4. The system has UV-lamp for classical epi-fluorescence observation, five lines of CW laser for confocal studies, femtosecond (fs) and picosecond (ps) pulsed laser for nonlinear microscope modalities. The fs laser source is a tunable, Ti:sapphire laser emitting around 690-1040 nm for efficient TPEF, or for higher SHG/THG spatial resolution.

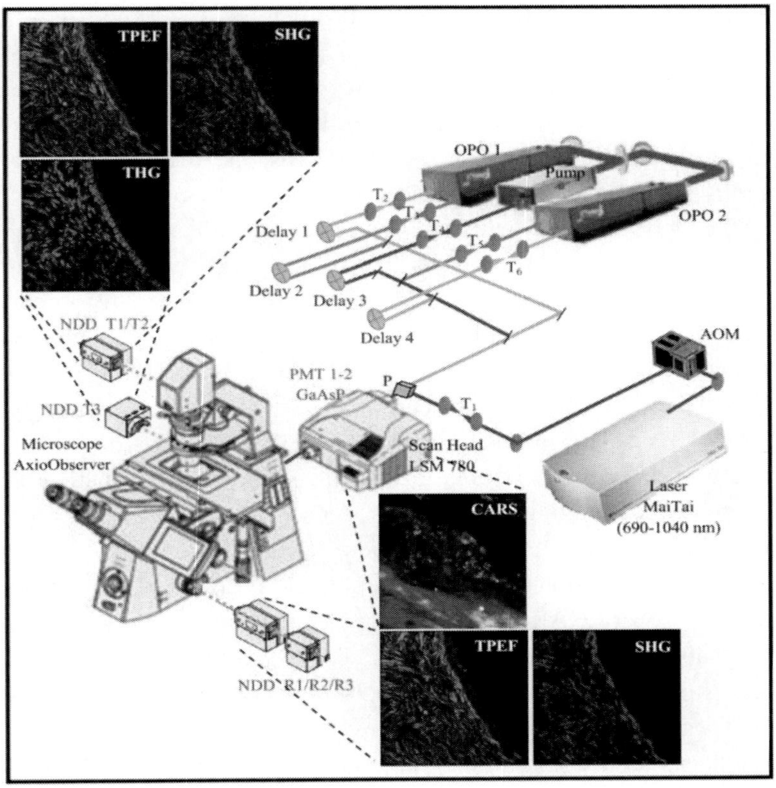

Figure 4. Schematic setup of a multimodal NLO microscope using a femtosecond laser source (Mai-Tai) and picoseconds (OPOs system) pulsed laser. Lasers are couple to on a commercial confocal system platform (inverted Zeiss Axio Observer.Z1 and confocal LSM 780). Principal optics elements are shown with blue letterings (telescope, delay lines, and recombination), and detectors with red letterings. Solid and dashed lines represent excitation and emission paths respectively. Representative TPEF (green), SHG (red), THG (magenta), and CARS (orange/yellow) ovarian images obtained with the setup are shown. Abbreviations: AOM-acousto-optic modulator, T: Telescope Lens, P: Recombine. NDD T: Non Descanned Detector (transmitted way), NDD R: Non Descanned Detector (reflective way), PMT: photomultiplier tubes. OPO: optical parametric oscillator, TPEF: two photon excited fluorescence, SHG: second harmonic generation, THG: third harmonic generation, CARS: coherent anti-Stokes Raman scattering.

The picoseconds (ps) source is obtained from a synchronously pumped optical parametric oscillator (OPO) system to obtain THG signal in the visible range and high spectral resolution CARS microscopy. The OPO can be easily and continuously tuned over a wide spectral range from 690 to 990 nm for the signal and between 1150 and 2450 nm for the idler output.

The fs beam is coupled to the scan head through an acousto-optic modulator (AOM) for a fast intensity control and with a collimating telescope (T_1) to adjust both the beam diameter to fill the objective back-aperture of the lens and the beam focus position on the microscope focal plane. The five wavelengths (signal and idler for each OPO plus the fundamental 1064 nm) are controlled independently with dedicated telescopes (T_2, T_3, T_4, T_5, and T_6). Delay lines on the five beam paths ensure temporal overlap between the beams. These beams necessary for CARS microscopy are temporally synchronized, recombined (P), and sent onto the backward excitation port of the scan head. The scan head of the LSM780 has a spectral GaAsP detector with 32 in-line elements and 2 adjacent PMTs. The motorized collimators, the scanners, the precisely positionable pinhole and the highly sensitive detectors are arranged to provide optimum specimen illumination and efficient collection of the emitted light.

The spectral width of a picosecond pulse is comparable to the Raman line width, so that the excitation energy is fully utilized to maximize the vibrationally resonant CARS signal. It has been shown theoretically that the optimal signal-to-background ratio occurs with spectral pulse widths of 1~3 ps for typical Raman band [11]. A pulse width of a few picoseconds provides a good compromise between the spectral resolution and the peak power, and improved the signal-to-background ratio. The OPO light source for CARS that we used has already been reported to perform time resolved CARS [47], to improve the sensitivity or to cover the full vibrational Raman spectrum. In figure 4 a possible setup is presented where the frequency doubled Nd :YVO generates 8 W of green light (532 nm) and pumps two OPOs with 4 W each. Although the full system can provide up to five wavelengths simultaneously (signal and idler for each OPO plus the fundamental 1064 nm), we normally use three colors. The two signals coming from each OPO are recombined with the fundamental 1064 nm beam from the Nd:YVO oscillator. The power, polarization state and divergence of each beam are controlled independently with dedicated polarization optics and telescopes. Delay lines on two of the beam paths (signal and 1064 nm) assure temporal overlap between the three beams. The backward detection is achieved with internal GaAsP detector of scan head. A set of dedicated filters are placed before the detectors to select the relevant spectral domains.

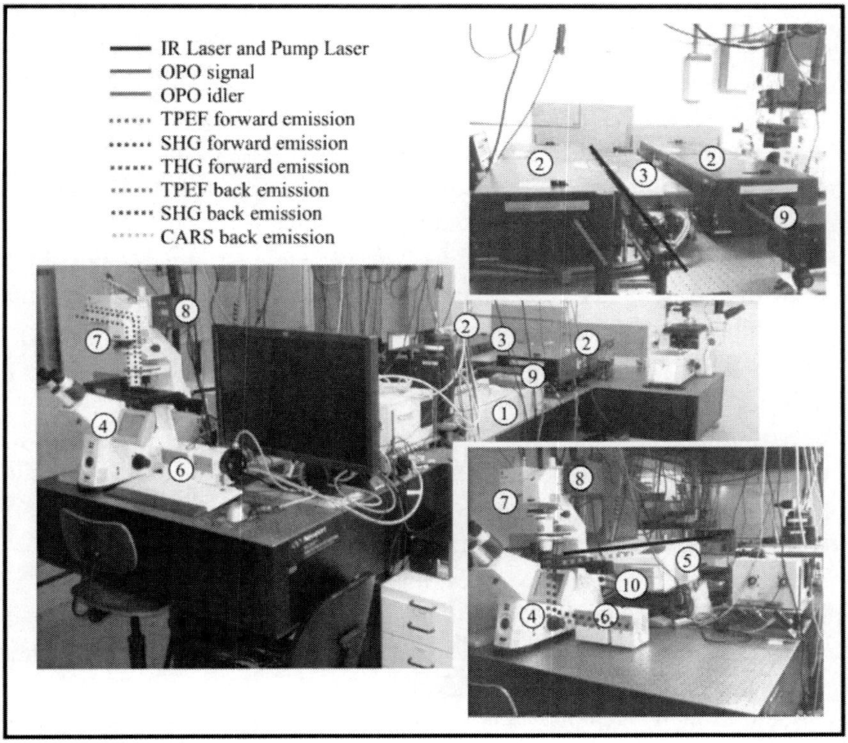

Figure 5. Real setup showing in the center multimodal platform (inverted Zeiss Axio Observer.Z1 and confocal LSM 780). Up panel show OPOs and Pump laser for CARS and down inverted microscope and scan head. Numbers: 1: Mai-Tai laser and DeepSee module, 2: OPOs, 3: Nd:YVO laser (fundamental), 4: Zeiss Axio Observer.Z1, 5: scan head, 6: NDD/R, 7: NDD/T, 8: holder UV lamp, 9: AOM, 10: GaAsP detector. Solid lines represent excitation path, and dashed line emission path. Abbreviations: IR: infrared, AOM: acousto-optic modulator, NDD/R: Non Descanned Detector (reflective way), NDD/T: Non Descanned Detector (transmitted way), GaAsP detector-Gallium Arsenite Phosphate, OPO: optical parametric oscillator, TPEF: two photon excited fluorescence, SHG: second harmonic generation, THG: third harmonic generation, CARS: coherent anti-Stokes Raman scattering.

The generated signals can be collected with the same microscope objective (reflected or epi-detected), splitted by dichroic mirrors, which were specifically chosen for the given fundamental and fluorescence or harmonic emission wavelengths and focused onto specific detectors. Interference or band pass filters are used in front of the detector for filtering scattered fundamental light and spurious signals outside the desired bandwidth. For epi-detection the system has two internal photomultiplier tubes (PMT) plus GaAsP avalanche photodiodes inside the scan head (PMT 1-2, figure 4). Also three Non Descanned Detectors (NND) are available for epi-detection of nonlinear signals (NDD R1/R2/R3). It is also possible to detect the signals in the forward direction. Either one detector with appropriate filters, or several detectors recording different signals separated by dichroic mirrors can be used. The system shown here has three NDD detectors (NDD T1/T2/T3). T3 detector was place right after the sample, holding it as close to the sample as possible.

With this system many configuration can be used and different, linear and/or nonlinear signals can be simultaneously detected. For example, we can observe CARS signal with internal GaAsP detector of scan head, forward THG, SHG and TPFE with NDD T1/T2/T3, and reflected SHG and TPEF with NDD R1/R2.

Figure 5 depicts the nonlinear microscope setup in the laboratory with a complete illustration of the different light paths, filters, and detectors distributions that compose this multimodal system.

Dark red lines indicate NIR excitation light from Ti:Sa or Nd:YVO oscillator lasers directed onto the sample for NLO modality. Green and like blue lines indicate signal and idler for each OPO. Green, red, and blue lines indicate backward o forward propagating TPEF, SHG, and THG signal respectively. Finally, yellow line represents backward CARS signal. All TPFE, SHG, THG, and CARS images, depicted in figure 4, were acquired with 1024 × 1024 pixel spatial resolution, using 3 μs for the pixel dwell time, with total scanning time of the order of 2s, after a 8 frames average. Images (pseudo-green TPEF, pseudo-red SHG, pseudo-magenta THG, and pseudo-yellow/red CARS) were combined for visualization using Zen 2009 Light Edition Image software.

III. Characterization of NLO Signals

As demonstration and example, the ability of the one setup system describe here was evaluate, to detect SHG and THG signal using a wavelength of the excitation below 1,000 nm. As was previously mentioned, 940 nm beam excitation generate a SHG signal at 470 nm (visible) and a THG signal at 313 nm (ultraviolet) respectively. To collect transmitted SHG and THG simultaneously we would need to use a high numerical aperture (NA) condenser [48, 49] to collect both signal and then split them with a dichroic to two PMTs. The condenser shall have a NA similar or higher than the objective lens in order to achieve optimal collection of the signals and also a good transmission curve in UV region [4]. However, our specific condenser had neither the right NA nor the good UV transmission, therefore we decided to simply remove the condenser and place the PMT as close as possible to the sample to increase the collection angle. For all laser powers tested (100 to 400 mW) good SHG and THG signal were detected using the PMT directly over the sample (without condenser lens). Therefore, has demonstrated that removing the condenser lens and using a special support for an external PMT (see figure 4, top photographs) it is possible to get good THG signal from a fs Ti:sapphire laser. However, in this geometry it was impossible to collect both, transmitted SHG and THG, simultaneously.

The THG contrast to local heterogeneities of size smaller or comparable to the beam focus was tested [28]. To do that, was analyzed the THG generated by the 940 nm laser beam in the air-glass interface where the Mai Tai laser was focused with a lens into a coverslip positioned in a translation stage. The resultant emission was registered using a monochromator/CCD system. THG signal only appeared, visible to a naked eye, when the laser was focused at the air-glass interface. It disappeared when the lens was removed but not at the 940 nm excitation, which was also detected by the CCD. The laser power was increased from 430mW to 1,140mW, measured before the lens, for different laser intensities. Figure 6A (black plot) shows the result of the wavelength integrated THG signal versus laser power in a log-log plot, which presented the expected THG signal proportional to the cube of the laser intensity. Similarly then was conducted an experiment directly on the microscope setup shown in figure 3 by placing a coverslip (170 μm thickness) in the objective focus and acquiring several scanning images at different depths (Figure 6B). The THG intensity was obtained with the Olympus Fluoview software. Figure 6C shows the plot of the THG intensity vs focal position which presents a peak at the focal position and disappeared after ~1 μm dislocation for each side. Next the filters were exchanged to detect the SHG signal from ovary tissues specimen which has rich collagen content in the stroma [50], a protein well known as a good source of SHG signal [51]. Using this sample the SHG intensity was measured as a function of laser power varying from 10 mW to 450 mW and verifying that it was proportional to the second power of the incident light (Figure 6A, red plot). The intensity value is an average of the whole image and was measured using ImageJ software. Taken all these results together were verified, therefore, that actually second and third harmonic generation signals were observed.

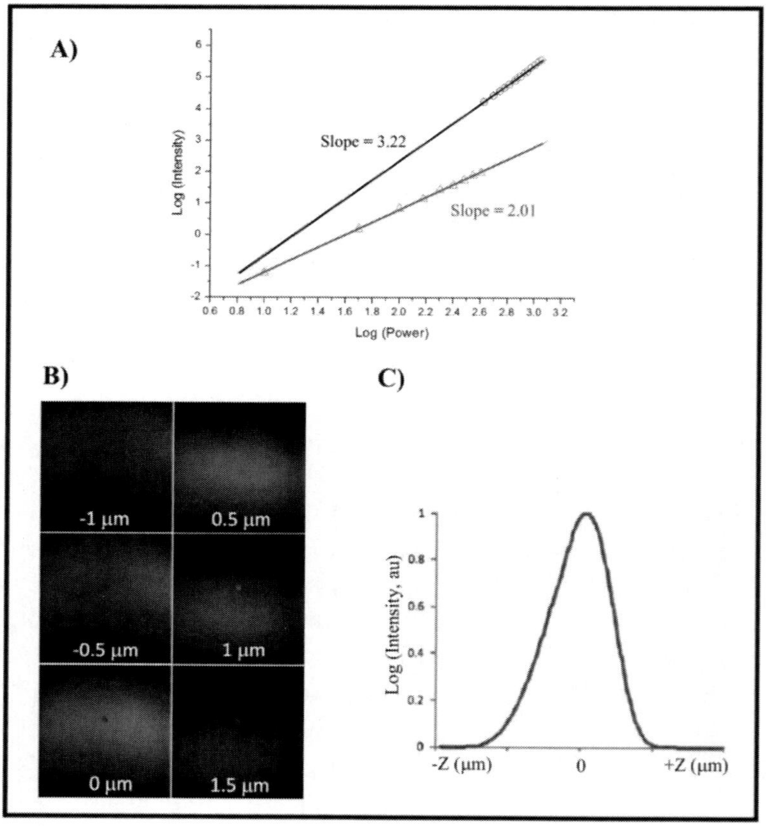

Figure 6. SHG and THG characterizations. A) Log-log graph of THG intensity (black plot) and SHG intensity (red plot) for laser output power. The circles and triangles are experimental data and the solid lines are lines with slopes near to 2 (red) and 3 (black), respectively. B) Several THG images at different depth of glass-air interface of coverslip. C) TGH intensity of signal shows in B, the maximum is obtained when the interface are in focal position.

CARS signal are more easily characterized and verified. Since to obtain this signal is needed the temporal overlap between the beams; will be sufficient to block one of them and check if the signal is detected. That can be obtained blocking the pump beam and corroborating that CARS signal disappeared, following unlocking this beam, the signal must be immediately recovered. Also, the Stokes beam blockage must cause CARS signal duck-out and also it must return, after unlocking.

IV. Applications for Epithelial Cancers Diagnosis

Cancers cannot be considered only in terms of their malignant cellular components, but also depending of the stromal environment which is altered to favor the cancer progression and invasion. The combination of different image approaches described in this chapter may represent a powerful combination of tools to study both, malignant cells and stromal environment. One of the main examples of such objective is the collagen organization changes analysis, the remodeling matrix and alterations in epithelial/stromal interface. Highly valuable, structural information revealed by each nonlinear contrast approach can be isolated and analyzed separately, while their superposition allows a better comparison and understanding of the spatial tissue organization. Thus, TPEF and THG can be used to image a variety of well-documented morphologic and architectural alterations, moreover combined TPEF-SHG can be applied to analyze alterations in epithelial cells and the supporting stroma, and CARS microscopy can be used to understand lipid and proteins composition in tumor tissues.

In the following section, some current NLO microscopy approaches in the cancer diagnosis are presented. The section starts with the description of structural and compositional images of normal tissues, followed by pathological examples. In particular, the application of these techniques to human breast, ovarian, and colon cancer and additionally in an induced colon cancer mouse model will be discussed. Finally, we will try to summarize the major milestones developed by other research groups over the last years in cancer diagnoses.

Figure 7 depicts representative H&E, TPEF, SHG, and THG images and their combinations of normal breast, ovary and colon tissues. The H&E-stained images shall be used as standard reference. TPEF signal (green) come from the elastin and collagen autofluorescence, which is weak in nuclear regions but strong in the regions outside the nucleus, revealing stromal connective structures. Characteristic microscopic appearance of each tissue type, and relationship between cells (THG) and stromal (SHG) can be identified. Further, the specific identification and characterization that can be achieved with the different NLO techniques; is also relevant that the image processing of this set of NLO microscopy approaches could make possible to reveal the relationship between signals of epithelial cells and the collagen matrix with epithelial carcinogenesis [8].

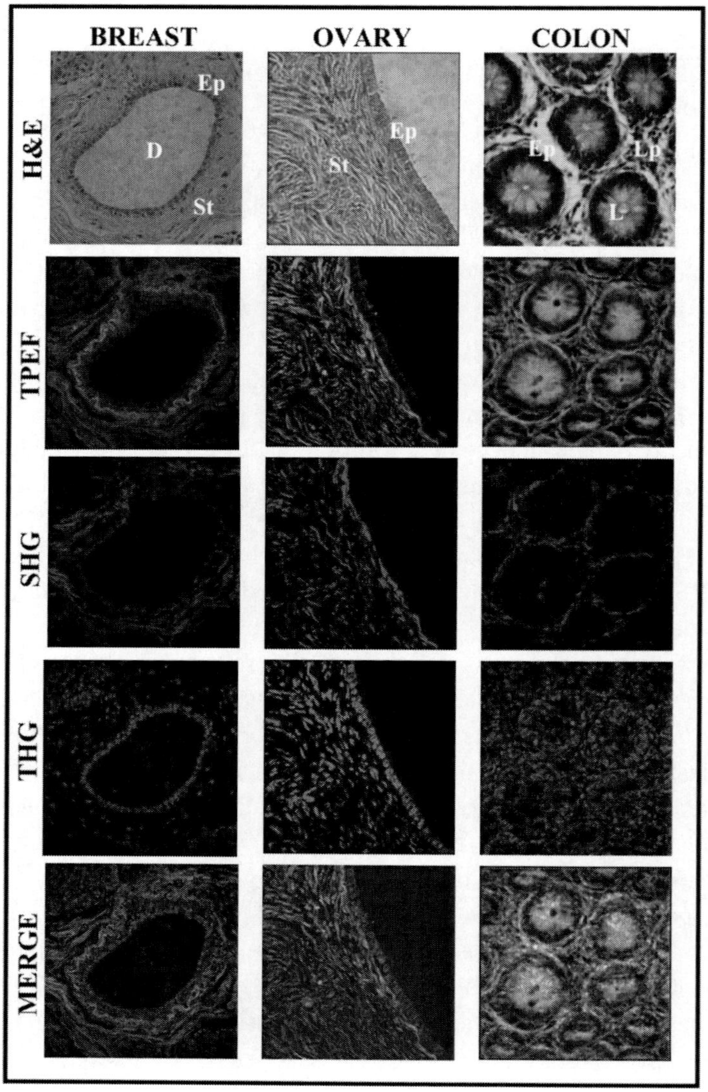

Figure 7. Representative H&E-stained and TPEF (green), SHG (red), THG (magenta), and (TPEF+SHG+THG) cross-sectional images of breast, ovary, and colon tissues diagnosed as normal. Abbreviations: Ep: epithelium, St: stroma, D: duct, L: luminal crypt, Lp: lamina propia.

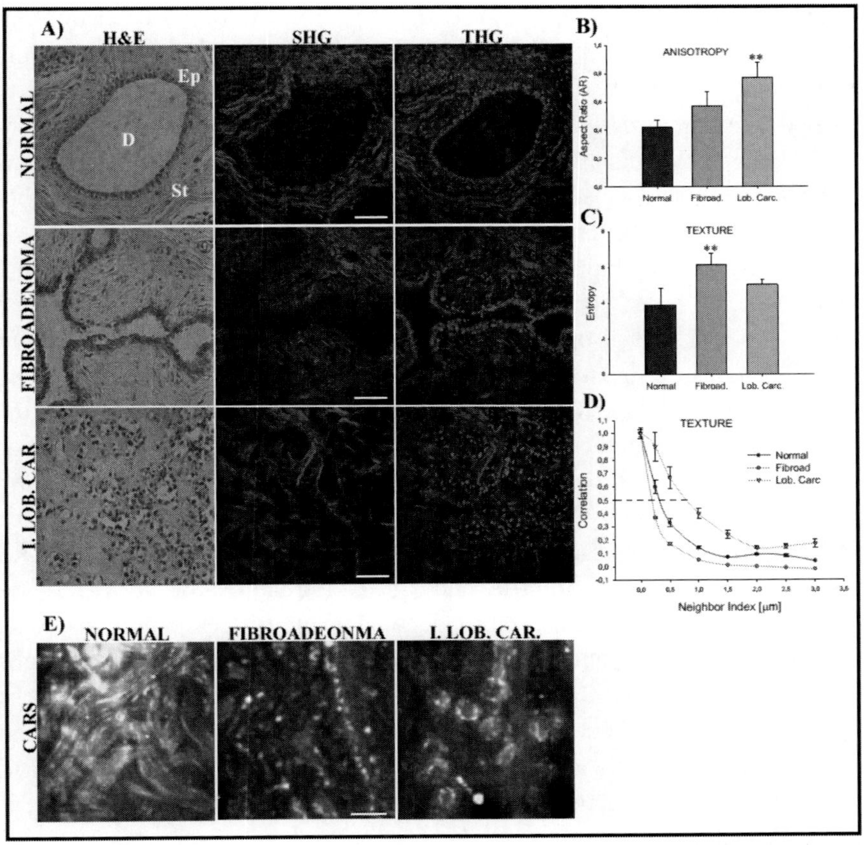

Figure 8. Multimodal NLO approach and scoring methods applied to human breast tumor. A) Representative H&E-stained and, SHG, and SHG+THG cross-sectional images of breast tissues diagnosed as normal (first row), fibroadenoma (second row), and invasive lobular carcinoma (third row). Scale bar = 20 µm. Bars graphs represent anisotropy (B aspect ratio) and texture quantitative analysis (C entropy and D correlation value). Each bar represents the mean ± SD of independent measurements. Asterisks indicate a very significant increase or decrease as compared to the non-tumor tissues (t-test, $p<0.01$). E) Representative CARS cross-sectional images of breast tissues. Scale bar = 20 µm. Abbreviations: D: duct, Ep: epithelium, St: stroma, Fibroad: fibroadenoma, Lob.Carc.: lobular carcinoma, I.Lob.Carc: Invasive lobular carcinoma.

Figure 8A exemplify this combination, where NLO techniques and quantification of collagen-related changes using anisotropy and texture features, were applied to differentiate between normal and malignant (Fibroadenoma and Invasive Lobular Carcinoma (ILC)) human breast tissue.

Characteristic microscopic appearance of each type of tissue, and relationship between cells and stromal can be identified, in the SHG/THG combination. The differential orientation and distribution of collagen fibers can be clearly identified in stroma region with SHG image. We evaluated these collagen-related changes by anisotropy (Figure 8B) and texture features calculations (Figure 8C, 8D). The aspect ratio (AR) value show that ILC present a higher AR (with significant difference), and fibroadenoma have an AR around 0.5, which means that they have a particular orientation (Figure 8B). As expected, malignant breast tissues present higher entropy that usually implies elevated levels of disorder and disorganization (Figure 8C). Moreover, correlation for the fibrils in ILC remained elevated for larger distances (between 0.5 and 1.5μm), implying less defined fibrillar structure. We have found that the correlation remained higher in ILC tissues with the $Corr_{50}$, the pixel distance where the correlation dropped below 50% of the initial value, significantly greater when compared with fibroadenoma and control tissues ($p<0.05$, ANOVA) (Figure 8D). Figure 8E shows a comparative analysis of CARS images of breast tissue. Adipose and fibrous structures of normal tissue possess strong CARS signals. Fibroadenoma exhibits the compressed duct with linear branching pattern whereas ILC present single or rows of cells invading into the stroma. Based on examples like this, and our previous works [35] we have established that it is possible to have both qualitative and quantitative parameters of differences between each kind of breast tumor and to demonstrate the advantage of the integration of as many NLO approaches as possible to analyze breast cancer.

Using these multimodal NLO approach and different scoring methods, we analyzed human ovarian tumors. Figure 9 shows a summary of the principal outcomes that allowed us to evaluate serous and mucinous ovarian tumors and also, to discriminate between subtypes such as adenoma, borderline and adenocarcinoma. First we shows combined TPEF+SHG+THG images of normal and serous and mucinous types ovarian tumors compared with classical H&E-stains sections (Figure 9A). TPEF signal depicts connective structures in the stroma. The SHG signal corresponds to collagen fibers within the connective tissue, while THG highlight the nuclei populations. Representative features of each technique are clearly distinguishable in the enlarged images (Figure 9A). The rich structural information revealed by each nonlinear contrast device can be directly compared, providing a better understanding about the tissue´s spatial organization. For example, SHG and THG signals provide a clear contrast to distinguish the epithelial/stromal interface (Figure 9C). It is worth mentioning that these differences and contrasts could be automatically and digitally done for quantification. For example, using merged TPEF+SHG signals, we have measured the stroma collagen/elastin content ratio (Figure 9B). Collagen content increased gradually within the stroma with tumor progression. The quantification of these observations was performed using SAAID (SHG to autofluorescence aging index of dermis) [52] which are showed by bar graph in the figure. The corresponding SAAID of both adenocarcinoma types presented statistically significant ($p<0.05$, t-test) higher values compared to normal stroma due to both the high SHG (collagen) and low TPEF signals in this region (Figure 9D). Other parameter of quantification useful was to measure (orientations and distribution of collagen fibers) by anisotropy performed on the SHG image of stromal region (Figure 9C). Anisotropy was calculated applying the Fourier transform analysis in specific region using ROI of 150x150 pixel side squared, to make sure that we are observing only the collagen network in the vicinity of the epithelium. In serous type tumors, AR increased progressively and significantly ($p<0.05$, ANOVA) from normal to adenocarcinoma and in mucinous type tumors (by contrast), AR showed statistically very significant differences only for adenocarcinomas ($p<0.01$, ANOVA) (Figure 9E). These results confirm the fact that, normal ovaries are more organized tissues as compared to adenocarcinoma. With this tool was possible to discriminate between serous adenoma from mucinous adenoma and serous borderline from mucinous borderline subtypes. Unlike ovarian serous tumors, which are relatively homogeneous in their cellular composition and differentiation degree, mucinous tumors are frequently heterogeneous, with mixtures of benign, borderline, and malignant elements often found within the same neoplasm. The heterogeneity in these mucinous tumors suggests that malignant transformation is sequential and slow, progressing from cystoadenoma to borderline tumor and, finally, to invasive carcinoma [53]. This slow behavior is probably reflected in a more organized stroma. Finally, using THG signal was possible to evaluate the differences in the surface epithelium of each tumor type (Figure 9C). In non-tumor sample, cells were arranged in a single cell layer. Serous adenoma samples show tall ciliated and non-ciliated cuboidal epithelial cells with elongated nuclei, also in one single cell layer with uniform distribution. On the other hand, the serous borderline tumor and serous adenocarcinoma samples are completely different from the previous ones, showing epithelial surface with cells of different sizes distributed in multiple layers, including cellular atypia and proliferation. Mucinous tumor samples are similar to borderline/adenocarcinoma, with cells of different sizes forming up multiple layers, but containing abundant cytoplasmic mucin and basal nuclei.

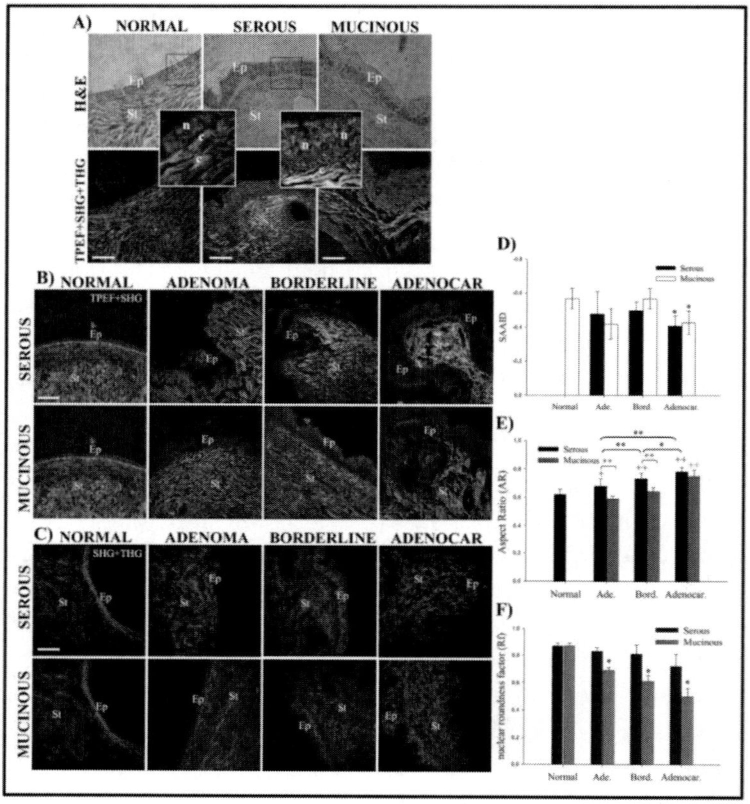

Figure 9. Multimodal NLO approach and scoring methods applied to human ovarian tumor. A) Representative H&E-stained and TPEF+SHG+THG cross-sectional images of ovarian tissues diagnosed as normal (first column), serous type tumor (second column), and mucinous type tumor (third row). B) Representative merges of TPEF and SHG cross-sectional images of ovarian tumor tissues. C) Representative merges of SHG and THG cross-sectional images of ovarian tumor tissues. D) Bars graphs represent SAAID index quantitative analysis. Asterisks indicate a significant increase as compared to the non-tumor tissues ($p<0.05$, t-test). E) Bars graphs represent anisotropy (aspect ratio) quantitative analysis. Comparisons with normal tissues are indicated with +. +,* indicates a statistically significant ($p<0.05$) difference and ++, ** indicates a statistically very significant ($p<0.01$) difference following ANOVA. F) Bars graphs represent quantitative analysis of the nucleus (roundness factor). Asterisks indicate a significant increase as compared to the non-tumor tissues ($p<0.05$, t-test). (A-C) All scale bar = 20 μm. (D-F) Each bar represents the mean ± SD of independent measurements. Abbreviations: Ep: epithelium, St: stroma, n: nucleus, c: collagen, Ade: adenoma, Bord: borderline, Adenocar: adenocarcinoma, SAAID: SHG to autofluorescence aging index of dermis.

The automatic quantitative evaluation of the nuclear roundness factor (Rf) and kurtosis (K) confirmed the qualitative characteristics described above. The mucinous Rf mean values are significantly different from the normal, while the serous did not exhibit significant differences. However, the borderline/adenocarcinoma Rf standard deviations are between 5 to 10 larger than the normal/adenoma ones, indicating a more pronounced irregularity of the tumor cell nuclei (Figure 9F). A complete description of the utility of the NLO techniques approaches applied to ovarian tumor could be found in [3, 36, 50, 54, 55].

Figure 10 summarizes the combined uses of these techniques in the analysis of human colon cancer. NLO images clearly demonstrate the circular arrangement pattern of control colonic crypts registered from crypt-cross sections, characterized by epithelial columnar cells and interspersed goblet cells (Figure 10A). TPEF revealed the typical foveolar pattern of colon mucosa glands, displaying crypts with rounded-luminal openings. SHG (red) specifically trace the collagen scaffold within lamina propria whereas THG (magenta) highlight the nuclei of the mucosa epithelial cells (Figure 10A). Evaluation of colonic tissue by SHG (Figure 10B) and THG (Figure 10C) microscopies rapid- and clearly allow differentiating between adenocarcinoma and normal tissue states. Whereas individual crypts were easily identified with TPEF, the interspersed connective tissue was detected via SHG of collagen. As was previously stated, in tumoral tissues, SHG images, highlights changes in the surrounding fibrous stroma. The quantification of collagen prevalence over the tumor tissues SHG images, results in a significant difference compared to the normal colon, demonstrating an increased collagen density in tumor stroma (Figure 10B). Even when the classical histological techniques allow calculating the nucleus- cytoplasm ratio, the THG images allow to selectively detecting just the nuclei, making easier and objective the digitalization process. Enlarged and irregular shaped tumor cells exhibit increased nucleus-to-cytoplasm ratio, and area is significantly higher (Figure 10C). Another key aspect of NLO techniques in relation to the classical H&E ones is that they have high potential to produce 3D reconstructions and stereological studies. Figure 10D shows three-dimensional representations that allow the visualization of indistinguishable features in classical two-dimensional procedures. Maximum projection 3D representation using THG images clearly demonstrated the adenocarcinoma architectural change, in which the circular arrangements of each colonic crypt and the spacing between cells have been lost. Also, an attenuation of goblet cell / crypt density and mucin depletion were easily distinguished.

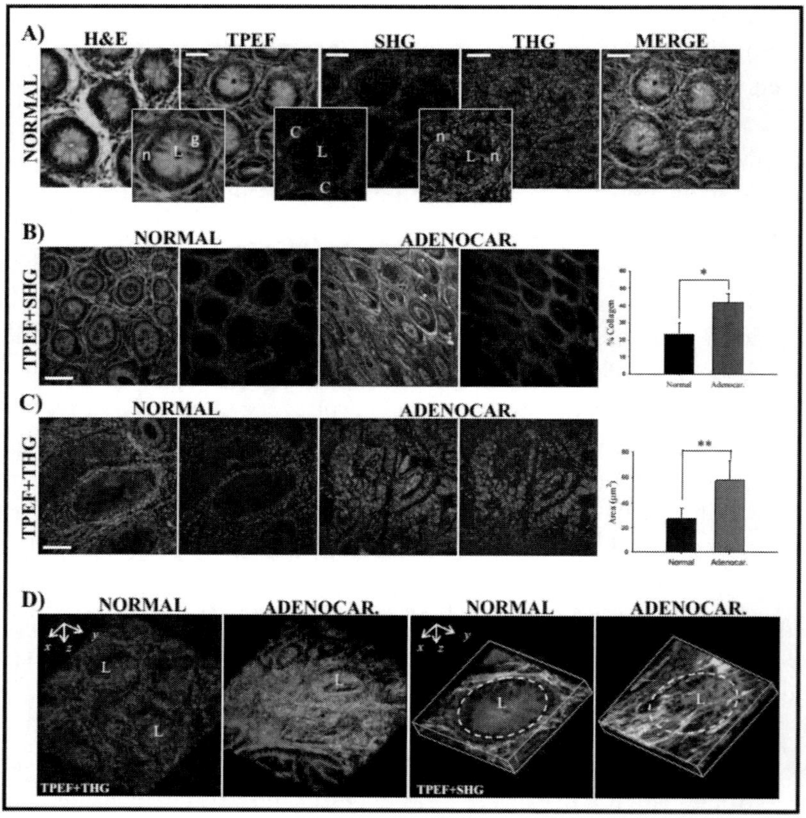

Figure 10. Multimodal NLO approach and scoring methods applied to human colon tumor. A) Representative H&E-stained and TPEF, SHG, THG, and MERGE cross-sectional images of colon tissues diagnosed as normal. B) Representative TPEF+SHG and SHG cross-sectional images of normal and tumor colon tissues. Bars graphs represent collagen density quantitative analysis. Asterisks indicate a significant increase as compared to the non-tumor tissues ($p<0.05$, t-test). C) Representative TPEF+THG and THG cross-sectional images of normal and tumor colon tissues. Bars graphs represent nucleus area quantitative analysis. Asterisks indicate a very significant increase as compared to the non-tumor tissues ($p<0.01$, t-test). D) Maximum projection of 60 images separated 0.5 μm each of normal and adenorcarcinomas colonic tissues. Epithelial-stromal interface is indicated (white outline). (B, C) Each bar represents the mean ± SD of independent measurements. Scale bars = 20 μm. Abbreviations: L: luminal crypt orifice, g: goblet cells, n: nucleus, c: collagen, Adenocar: adenocarcinoma.

Moreover, using a three-dimensional SHG representation it was possible to observe the inclination and invasiveness of collagen fibers of adenocarcinomas compared to normal colon. It is worth pointing out that these features (presence and angle of collagen fibers) are not easily visible in classical two-dimensional H&E-stained sections.

In conclusion, the use of several NLO microscopy approaches, especially when combined, can reveal information not distinguishable in H&E stained sections under classical bright field approaches. Cell nucleus size and shape, tightness and disposition changes of collagen fibers are parameters that can be consistently quantified, allowing predict a huge clinical potential in cancer colon. These results clearly show significant changes of collagen fiber morphology, orientation and density in colon tumor tissue, suggesting that collagen fiber inclination angles are an important factor in tumor progression. In agreement with these results, previous reports in human colon and other tissues suggest that the epithelial cells preferentially invade tissues where the collagen fibers became perpendicularly aligned, instead of randomly arranged ones [56, 57].

The recognition of premature cancer injury is an issue without solution. In this sense, non-invasive approaches show potential for cancer detection. The properties of NLO images described in this chapter could become, in the near future, in a valuable tool to detect abnormal mucosa and stroma configurations in early stages of the disease. Recently, working with mouse CCR model (in which the intestinal adenocarcinomas grow rapidly and their histological transformations remember those encountered in humans), we observed that SHG signals from collagen fibrils during tumor progression shows differences that might serve as an early biomarker of colon cancer risk.

Figure 11A, shows representative pictures of the induced CCR mouse model. The control mouse colon exhibits a crypt-architecture and an ordered cell polarity pattern. At week 4 of AOM/DSS treatment, start to appear colon lesions characteristics of the first pathological steps, the aberrant crypt foci (ACF). Eight weeks after injection, lesions become highly dysplasic. The dysplastic epithelium contains lengthened, packed and dark-staining nuclei. The crypts architecture have been lost, depicting on the right of the image fused and round figures and on the left, long and slender villi, with tapered tips (Figure 11B). The combination of TPEF+SHG images enables a superior evaluation of the spatial organization of the tissue (Figure 11C, 11D). Using SHG signal, the amount and distribution of the collagen fibers were quantified at different weeks post AOM injection. In crypts, no differences in SHG signal were detected (Fig 11C) but in submucosa, the amount of collagen fibers was augmented in proportion with the pathology development (Figure 11D). Bars graphs depict the average of integrated density (ID) from all examined samples.

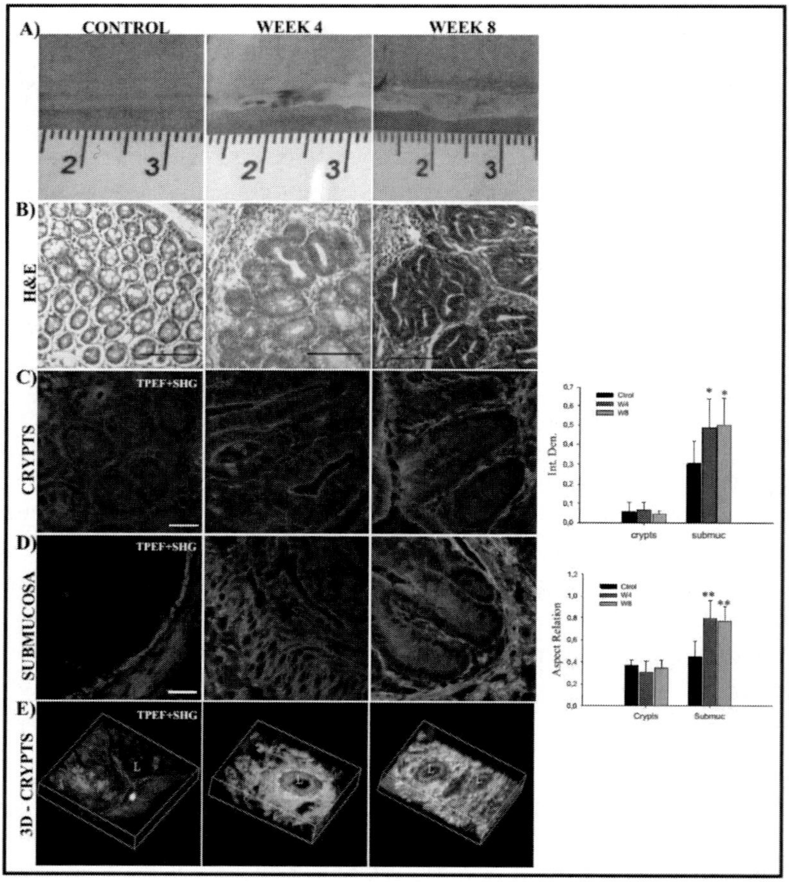

Figure 11. Multimodal NLO approach and scoring methods applied of colorectal cancer (CRC) mouse model. A) Colon cancer development after AOM injection. Macroscopic view of murine colon segments and B) H&E stained sections taken from (A) particular regions. Scale bars: 30 μm. C, D) TPEF and SHG images of mouse colon, crypts (C) and submucosa (D). Bars graphs represent the integrated density and aspect ratio quantitative analysis of colon samples. Asterisks indicate a significant o very significant increase as compared to the non-tumor tissues ($p<0.05$ - $p<0.01$, t-test). Scale bars: 50 μm. E) Maximum projection (TPEF+SHG) of individual crypts. 46 images separated 1 μm each. Abbreviations: L: luminal crypt orifice, Submuc: submucosa, w4: week 4, w8: week 8, ID: integrated density.

At weeks 4 and 8 post AOM injection ID was significantly higher ($p < 0.05$) compared with control samples (Figure 11C). Regarding the arrangement of the collagen fibers, anisotropy calculations were estimated. FFT images of control sample exhibit a more elliptical profile (fibers are arranged in parallel form) when compared to the FFT images of samples process with AOM, which presents a more circular configuration (fiber without any specific arrangement). The figure 11D shows the averaged AR of the examined stages. Again, at weeks 4 and 8, AR values were increased and extremely significant ($p < 0.01$), comparing to normal colon. These data allow verifying that normal submucosa is a more ordered tissue than the transformed one. The tridimensional representations (Figure 11E) bear out previous observation that collagen fibrils adopt a distinctive behavior relative to epithelium, as the pathology progresses. In conclusion, these preliminary results allow to visualize that this methodology could evolve as a non-invasive alternative for early colon cancer detection in a near future [58].

These techniques alone or combined in different platforms with multiple imaging processing and quantification tools have proven to be extremely useful for the study of cancer. Some reports in human studies of cancer showed that the NLO techniques presented sensitivity, specificity and positive predictive values of 90%, 80%, 95% respectively [59,60], which means that NLO techniques represent excellent detection methodology. Here we mentioned some specific studies performed in the last five years that confirm previous affirmation. Label-free nonlinear imaging (TPEF, SHG, and CARS) have demonstrated excellent correspondence with staining histopathology in case of renal and oral squamous cell carcinoma (SCC). Tumor boundary was clearly visible [41]. Combined SHG and THG imaging enables detect many architectural and cytological features characteristic for SCC [61]. Other studies have analyzed disease induced alterations in melanoma [62], basal and skin squamous cell carcinoma [63]. The categorization of normal and malignant breast tissue was realized using either CARS or SHG. [38, 64]. Visualization of the epithelium and differences in tissue architecture and the cellular appearance occurring during carcinogenesis are detectable for colonic cancer [65], head and neck squamous cell carcinoma [35, 61, 66, 67], kidney tumors [41], pancreas and pancreatic cancer [42, 68], bladder [60], ovary and ovarian tumors [3, 36, 54], and lung carcinoma [69]. Also the differentiation of benign, and malignant brain tissue [70-72] were performed.

CONCLUSION

In cancer diagnosis, there is a need for development of a multimodal imaging-based diagnostic tool to objectively evaluate morphological features with subcellular resolution and molecular compositions that are closely associated with tumor malignancy. With this perspective, NLO microscopy has proven to be useful in cancer research. These techniques have recently emerged as a valuable tool for high-resolution, nondestructive, chronic imaging of living tumors. Moreover, multi-modal microscopy can provide a powerful tool for investigating the dynamics of structure-function relationships at the molecular and subcellular levels.

In combination with sophisticated animal models and computer-assisted data analysis, NLO microscopy techniques and image processing methods are opening new doors to the study of tumor biology and facilitating the development of new strategies for early tumor diagnosis and treatment. Today, the application of multimodal nonlinear imaging is recognized in basic research in the biological and biomedical sciences, however regular applications in clinics are still rare, mainly because of their high cost. Multimodal platform are still complex, and require specialized personnel for its operation.

Therefore, especially technological improvements are required for miniaturization, improvement of the ease of handling and automated data processing and extraction of relevant information. An important step toward this goal is the modification of standard clinical endoscopes for multimodal nonlinear imaging suited for in vivo applications.

The development of nonlinear optical endoscopy, which allows imaging under conditions in which a conventional nonlinear optical microscope cannot be used, will be the primary goal to extend applications of nonlinear optical microscopy toward clinical ones. There are several key challenges involved in the pursuit of in vivo nonlinear optical endoscopy. Between them, can be mentioned the necessity of obtain efficient ultra-short pulse laser liberations into a remote place, the need to enhance scan rates for monitoring biological processes and also, the miniaturization of the laser-scanning mechanisms to the millimeter scale. Finally, the design of a nonlinear optical endoscope based on micro-optics with great flexibility, and compact enough to be incorporated into endoscopes, will become an evolution of these microscopy approaches.

These and other developing nonlinear applications, together with the ongoing progress in human in vivo NLO cancer imaging discussed in this chapter, will advance our ability to non-invasively diagnose and treat cancer into the coming years.

REFERENCES

[1] Masters BR and So PTC. The genesis of nonlineal microscopies and their impact on modern developments. In: Master and So Editors. Handbook of biomedical nonlinear optical microscopy. Oxford. University Press 2008, 5-28.

[2] Wilder-Smith P, Krasieva T, Jung WG, Zhang J, Chen Z, Osann K, Tromberg B. Noninvasive imaging of oral premalignancy and malignancy. *J. Biomed. Opt.* 2005;10(5):051601.

[3] Adur J, Pelegati VB, Costa LF, Pietro L, de Thomaz AA, Almeida DB, Bottcher-Luiz F, Andrade LA, Cesar CL. Recognition of serous ovarian tumors in human samples by multimodal nonlinear optical microscopy. *J. Biomed. Opt.* 2011;16(9):096017.

[4] Carriles R, Schafer DN, Sheetz KE, Field JJ, Cisek R, Barzda V, Sylvester AW, Squier JA. Imaging techniques for harmonic and multiphoton absorption fluorescence microscopy. *Rev. Sci. Instrum.* 2009; doi: 10.1063.

[5] Pelegati VB, Adur J, De Thomaz AA, Almeida DB, Baratti MO, Andrade LA, Bottcher-Luiz F, Cesar CL. Harmonic optical microscopy and fluorescence lifetime imaging platform for multimodal imaging. *Microsc Res. Tech.* 2012; 75 doi: 10.1002.

[6] Sun CK. Higher harmonic generation microscopy. *Adv. Biochem. Eng/Biotechnol.* 2005; 95:17-56.

[7] Meyer T, Schmitt M, Dietzek B, Popp J. Accumulating advantages, reducing limitations: multimodal nonlinear imaging in biomedical sciences - the synergy of multiple contrast mechanisms. *J. Biophotonics.* 2013; 6:887-904.

[8] Adur J, Carvalho HF, Cesar CL, Casco VH. Nonlinear optical microscopy signal processing strategies in cancer. *Cancer Inform.* 2014; 13:67-76.

[9] Provenzano PP, Eliceiri KW, Yan L, Ada-Nguema A, Conklin MW, Inman DR, Keely PJ. Nonlinear optical imaging of cellular processes in breast cancer. *Microsc Microanal.* 2008 Dec;14(6):532-48.

[10] Perry SW, Burke RM, Brown EB. Two-photon and second harmonic microscopy in clinical and translational cancer research. *Ann. Biomed. Eng.* 2012; 40:277-291.

[11] Cheng JX, Xie XS. Coherent Anti-Stokes Raman scattering microscopy: instrumentation, theory, and applications. *J. Phys. Chem B.* 2004; 108:827-840.

[12] Rodriguez LG1, Lockett SJ, Holtom GR. Coherent anti-stokes Raman scattering microscopy: a biological review. *Cytometry A.* 2006; 69:779-791.

[13] Shen YR. The Principles of Nonlineal Optics. New York, John Wiley, 1984.

[14] Gu M, Bird D, Day D, Fu L, Morrish D. Nonlinear optical microscopy. In Femtosecond Biophotonics Core Technology and Applications. Cambridge University Press 2010, 9-34.

[15] Denk W, Strickler JH, Webb WW. 1990. Two-photon laser scanning fluorescence microscopy. *Science* 248:73-76.

[16] Larson DR, Zipfel WR, Williams RM, Clark SW, Bruchez MP, Wise FW, Webb WW. Water-soluble quantum dots for multiphoton fluorescence imaging in vivo. *Science.* 2003; 300:1434-1436.

[17] Li J, Day D, Gu M. Ultra-low energy threshold for cancer photothermal therapy using transferrin-conjuagted gold nanorods. *Adv. Mater.* 2008; 20:1-6.

[18] Cheng PC, Sun CK. Nonlinear (Harmonic Generation) Optical Microscopy. In Handbook of Biological Confocal Microscopy. Third Edition. Pawley JB editor 2006, 703-721.

[19] Freund I, Deutsch M, Sprecher A. Connective tissue polarity. Optical second-harmonic microscopy, crossed-beam summation, and small-angle scattering in rat-tail tendon. *Biophys. J.* 1986; 50:693-712.

[20] Guo Y, Ho PP, Tirksliunas A, Liu F, Alfano RR. Optical harmonic generation from animal tissues by the use of picosecond and femtosecond laser pulses. *Appl. Opt.* 1996; 35:6810-6813.

[21] Guo Y, Ho PP, Savage H, Harris D, Sacks P, Schantz S, Liu F, Zhadin N, Alfano RR. Second harmonic tomography of tissues. *Opt. Lett.* 1997; 22:1323-1325.

[22] Guo Y, Savage HE, Liu F, Schantz SP, Ho PP, Alfano RR. Subsurface tumor progression investigated by noninvasive optical second harmonic tomography. *Proc. Natl. Acad. Sci.* USA. 1999; 96:10854-10856.

[23] Alfano RR, Guo Y, Liu F, Ho PP. Non-linear optical tomography of turbid media, US Patent #6,208,886 B1, (March 27, 2001).

[24] Campagnola PJ, Wei MD, Lewis A, Loew LM. High-resolution nonlinear optical imaging of live cells by second harmonic generation. *Biophys. J.* 1999; 77:3341-3349.

[25] Barad Y, Eisenberg H, Horowitz M, Silberberg Y. Nonlinear scanning laser microscopy by third harmonic generation. *Appl. Phys. Lett.* 1997; 70:922-924.

[26] Chu SW, Chen IH, Liu TM, Chen PC, Sun CK, Lin BL. Multimodal nonlinear spectral microscopy based on a femtosecond Cr:forsterite laser. *Opt. Lett.* 2001; 26:1909-1911.

[27] Boyd R. Nonlinear optics, 1st edn. Academic Press, New York. 1992

[28] Debarre D and Beaurepaire E. Quantitative characterization of biological liquids for third harmonic generation microscopy. *Biophys. J.* 2007; 92:603-612.

[29] Williams RM, Zipfel WR, Webb WW. Interpreting second-harmonic generation images of collagen I fibrils. *Biophys. J.* 2005; 88:1377-1386.

[30] Campagnola PJ, Millard AC, Terasaki M, Hoppe PE, Malone CJ, Mohler WA. Three-dimensional high-resolution second-harmonic generation imaging of endogenous structural proteins in biological tissues. *Biophys. J.* 2002; 82:493-508.

[31] Chu SW, Chen SY, Chern GW, Tsai TH, Chen YC, Lin BL, Sun CK. Studies of chi(2)/chi(3) tensors in submicron-scaled bio-tissues by polarization harmonics optical microscopy. *Biophys. J.* 2004; 86:3914-3922.

[32] Müller M, Squier J, Wilson KR, Brakenhoff GJ. 3D microscopy of transparent objects using third-harmonic generation. *J. Microsc.* 1998; 3:266-274.

[33] Moreaux L, Sandre O, Blanchard-Desce M, Mertz J. Membrane imaging by simultaneous second-harmonic generation and two-photon microscopy. *Opt. Lett.* 2000; 25:320-322.

[34] Sun CK, Chu SW, Chen SY, Tsai TH, Liu TM, Lin CY, Tsai HJ. Higher harmonic generation microscopy for developmental biology. *J. Struct. Biol.* 2004; 147:19-30.

[35] Adur J, Pelegati VB, de Thomaz AA, D'Souza-Li L, Assunção Mdo C, Bottcher-Luiz F, Andrade LA, Cesar CL. Quantitative changes in human epithelial cancers and osteogenesis imperfecta disease detected using nonlinear multicontrast microscopy. *J. Biomed. Opt.* 2012; 17:081407-1.

[36] Adur J, Pelegati VB, de Thomaz AA, Baratti MO, Andrade LA, Carvalho HF, Bottcher-Luiz F, Cesar CL. Second harmonic generation

microscopy as a powerful diagnostic imaging modality for human ovarian cancer. *J. Biophotonics.* 2014; 7:37-48.

[37] Duncan MD, Reintjes J, Manuccia TJ. Scanning coherent anti-Stokes Raman microscope. *Opt. Lett.* 1982; 7:350-352.

[38] Yang Y, Li F, Gao L, Wang Z, Thrall MJ, Shen SS, Wong KK, Wong ST. Differential diagnosis of breast cancer using quantitative, label-free and molecular vibrational imaging. *Biomed. Opt. Express.* 2011; 2:2160-2174.

[39] Le TT, Huff TB, Cheng JX. Coherent anti-Stokes Raman scattering imaging of lipids in cancer metastasis. *BMC Cancer.* 2009; 9 doi: 10.1186.

[40] Meyer T1, Bergner N, Bielecki C, Krafft C, Akimov D, Romeike BF, Reichart R, Kalff R, Dietzek B, Popp J. Nonlinear microscopy, infrared, and Raman microspectroscopy for brain tumor analysis. *J. Biomed. Opt.* 2011; 16 doi: 10.1117.

[41] Galli R, Sablinskas V, Dasevicius D, Laurinavicius A, Jankevicius F, Koch E, Steiner G. Non-linear optical microscopy of kidney tumours. *J Biophotonics.* 2014; 7(1-2):23-27.

[42] Hu W, Zhao G, Wang C, Zhang J, Fu L. Nonlinear optical microscopy for histology of fresh normal and cancerous pancreatic tissues. *PLoS One.* 2012; 7:e37962.

[43] Hompland T, Erikson A, Lindgren M, Lindmo T, de Lange Davies C. Second-harmonic generation in collagen as a potential cancer diagnostic parameter. *J. Biomed. Opt.* 2008; 13 doi: 10.1117.

[44] Sivaguru M, Durgam S, Ambekar R, Luedtke D, Fried G, Stewart A, Toussaint KC Jr. Quantitative analysis of collagen fiber organization in injured tendons using Fourier transform-second harmonic generation imaging. Opt Express. 2010; 18:24983-24993.

[45] Mouras R, Rischitor G, Downes A, Salter D, Elfick A. Nonlinear optical microscopy for drug delivery monitoring and cancer tissue imaging. *J. Raman Spectrosc.* 2010; 41:848-852.

[46] Bianchini P, Diaspro A. Three-dimensional (3D) backward and forward second harmonic generation (SHG) microscopy of biological tissues. *J. Biophotonics.* 2008; 1:443-50.

[47] Volkmer A, Book LD, Xie XS. Time-resolved coherent antiStokes Raman scattering microscopy: imaging based on Raman free induction decay. *App. Phys. Lett.* 2002; 80: 1505-1507.

[48] Millard AC, Campagnola PJ, Mohler W, Lewis A, Loew LM. Biophotonics, Part B, Methods in Enzymology Academic, New York,Washington, D.C., 2003; Vol. 361, pp. 47–69.

[49] Cox G, Moreno N, Feijó J. Second-harmonic imaging of plant polysaccharides. *J. Biomed. Opt.* 2005; 10:024013.

[50] Kirkpatrick ND, Brewer MA, Utzinger U. Endogenous optical biomarkers of ovarian cancer evaluated with multiphoton microscopy. *Cancer Epidemiol. Biomarkers Prev.* 2007; 16:2048-2057.

[51] Campagnola1 PJ, Yuan Dong C. Second harmonic generation microscopy: principles and applications to disease diagnosis. *Laser Photon. Rev.* 2011; 5:13-26.

[52] Cicchi R, Massi D, Sestini S, Carli P, De Giorgi V, Lotti T, Pavone FS. Multidimensional non-linear laser imaging of Basal Cell Carcinoma. *Opt. Express.* 2007; 15:10135-10148.

[53] Hart WR. Mucinous tumors of the ovary: a review. *Int. J. Gynecol. Pathol* (2005); 24:4–25.

[54] Adur J, Pelegati VB, de Thomaz AA, Baratti MO, Almeida DB, Andrade LA, Bottcher-Luiz F, Carvalho HF, Cesar CL. Optical biomarkers of serous and mucinous human ovarian tumor assessed with nonlinear optics microscopies. *PLoS One*. 2012; 7(10):e47007.

[55] Nadiarnykh O, LaComb RB, Brewer MA, Campagnola PJ. Alterations of the extracellular matrix in ovarian cancer studied by Second Harmonic Generation imaging microscopy. *BMC Cancer.* 2010; 10:94.doi:10.1186.

[56] Zhuo S, Zhu X, Wu G, Chen J, Xie S. Quantitative biomarkers of colonic dysplasia based on intrinsic second-harmonic generation signal. *J. Biomed. Opt.* 2011; 16:120501.

[57] Birk JW, Tadros M, Moezardalan K, Nadyarnykh O, Forouhar F, Anderson J, Campagnola P. Second Harmonic Generation Imaging Distinguishes Both High-Grade Dysplasia and Cancer from Normal Colonic Mucosa. *Dig. Dis. Sci.* 2014; doi: 10.1007.

[58] Bianchi M, Adur J, Ruff SY, Izaguirre MF, Carvalho HF, Cesar CL, Casco VH. Mouse colorectal cancer an early detection approach using nonlinear microscopy. *Bio-Med. Mat. and Eng.* 2014; (In press).

[59] Yan J, Zhuo S, Chen G, Wu X, Zhou D, Xie S, Jiang J, Ying M, Jia F, Chen J, Zhou J. Preclinical study of using multiphoton microscopy to diagnose liver cancer and differentiate benign and malignant liver lesions. *J. Biomed. Opt.* 2012; 2:026004 doi: 10.1117.

[60] Jain M, Robinson BD, Scherr DS, Sterling J, Lee MM, Wysock J, Rubin MA, Maxfield FR, Zipfel WR, Webb WW, Mukherjee S. Multiphoton microscopy in the evaluation of human bladder biopsies. *Arch. Pathol. Lab. Med.* 2012; 136:517-526.

[61] Tsai MR, Shieh DB, Lou PJ, Lin CF, Sun CK. Characterization of oral squamous cell carcinoma based on higher-harmonic generation microscopy. *J. Biophotonics.* 2012; (5-6):415-424.

[62] Thrasivoulou C, Virich G, Krenacs T, Korom I, Becker DL. Optical delineation of human malignant melanoma using second harmonic imaging of collagen. *Biomed. Opt. Express.* 2011; 2:1282-95. doi: 10.1364.

[63] Heuke S, Vogler N, Meyer T, Akimov D, Kluschke F, Röwert-Huber HJ, Lademann J, Dietzek B, Popp J. Multimodal mapping of human skin. *Br. J. Dermatol.* 2013; 169:794-803.

[64] Ambekar R, Lau TY, Walsh M, Bhargava R, Toussaint KC Jr. Quantifying collagen structure in breast biopsies using second-harmonic generation imaging. *Biomed. Opt. Express.* 2012; 3:2021-2035.

[65] Zhuo S, Yan J, Chen G, Chen J, Liu Y, Lu J, Zhu X, Jiang X, Xie S. Label-free monitoring of colonic cancer progression using multiphoton microscopy. *Biomed. Opt. Express.* 2011; 2:615-619.

[66] Meyer T, Guntinas-Lichius O, von Eggeling F, Ernst G, Akimov D, Schmitt M, Dietzek B, Popp J. Multimodal nonlinear microscopic investigations on head and neck squamous cell carcinoma: toward intraoperative imaging. *Head Neck.* 2013; 35:E280-E287.

[67] Teh SK, Zheng W, Li S, Li D, Zeng Y, Yang Y, Qu JY. Multimodal nonlinear optical microscopy improves the accuracy of early diagnosis of squamous intraepithelial neoplasia. *J. Biomed. Opt.* 2013; 18:036001.

[68] Hu W, Fu L. Simultaneous characterization of pancreatic stellate cells and other pancreatic components within three-dimensional tissue environment during chronic pancreatitis. *J. Biomed. Opt.* 2013; 18:56002.

[69] Gao L, Wang Z, Li F, Hammoudi AA, Thrall MJ, Cagle PT, Wong ST. Differential diagnosis of lung carcinoma with coherent anti-Stokes Raman scattering imaging. *Arch. Pathol. Lab. Med.* 2012; 136:1502-1510.

[70] Freudiger CW, Pfannl R, Orringer DA, Saar BG, Ji M, Zeng Q, Ottoboni L, Wei Y, Waeber C, Sims JR, De Jager PL, Sagher O, Philbert MA, Xu X, Kesari S, Xie XS, Young GS. Multicolored stain-free histopathology with coherent Raman imaging. *Lab. Invest.* 2012; 92:1492-1502.

[71] Ji M1, Orringer DA, Freudiger CW, Ramkissoon S, Liu X, Lau D, Golby AJ, Norton I, Hayashi M, Agar NY, Young GS, Spino C, Santagata S, Camelo-Piragua S,Ligon KL, Sagher O, Xie XS. Rapid, label-free detection of brain tumors with stimulated Raman scattering microscopy. *Sci. Transl. Med.* 2013; 5:201ra119.

[72] Meyer T1, Bergner N, Bielecki C, Krafft C, Akimov D, Romeike BF, Reichart R, Kalff R, Dietzek B, Popp J. Nonlinear microscopy, infrared, and Raman microspectroscopy for brain tumor analysis. *J. Biomed. Opt.* 2011;16:021113.

In: Nonlinear Optics
Editor: Fred Wilkins

ISBN: 978-1-63321-926-7

Chapter 4

NONLINEAR OPTICS IN OPTICAL SLOT-MICROFIBER

Fei Xu*, Wei Luo, Wei Guo, Wei Hu, Yan-qing Lu, Zhen-xing Wu and Jun-long Kou

College of Engineering and Applied Sciences and National Laboratory of Solid State Microstructures, Nanjing University, Nanjing, P. R. China

ABSTRACT

Optical microfibers (MFs) are fabricated by adiabatically stretching conventional optical fiber thus preserving the original optical fiber dimensions at their input/output pigtails, allowing ready splicing to standard fibers. MFs diameters, close to the wavelength of propagating light, are of interest for a range of emerging fiber optic applications since they offer a number of enabling optical and mechanical properties including large evanescent fields, high nonlinearity, extreme flexibility and configurability, low-loss interconnection to other optical fibers and fiberized components. However, its application will still be limited because MFs are generally circular symmetric. Here we review one kind of novel MF with non-circular symmetric geometry, the fiberized slot waveguides (FSW) or slot-microfiber (slot-MF), which is a highly birefringent engineered MF with nano-scale slots where light field can be confined and enhanced. The slot-MF has the advantages of both a MF and a slot waveguide. Taking advantage of the unique and flexible

* Corresponding author's email: feixu@nju.edu.cn

geometry, a number of significant linear and nonlinear characteristics can be realized including high birefringence, ultra-flattened and low dispersion profiles, and strong surface second-order nonlinearity. With the incorporation of high-index fiber and high nonlinearity material inside the slot, slot-MFs should be ideally suited for a host of nonlinear optical interactions and devices, and offers a large potential in miniature fiber devices for all-optical, fast operating functions for optical telecommunications.

Keywords: microfiber, slot, surface second harmonics, nonlinear, dispersion

1. INTRODUCTION

In recent years, there is increasing interest in the research of optical fiber nanowires or microfibers (MFs), which are fibers with submicrometer- and nanometer- diameters [1]. The size is tens to thousands of times thinner than the standard optical fiber. With the advantages of large evanescent fields, high nonlinearity, extreme flexibility and configurability and low-loss interconnection to other optical fibers and fiberized components, they have found potential applications in a wide range of fields of optical communications, sensing, lasers, biology and chemistry [1-4]. MFs are fabricated by adiabatically stretching optical fibers while keeping the original dimensions of the optical fiber at their input and output allowing ready splicing to standard fibers. Generally, the cross-section is circular-symmetric, and the MF is nearly zero-birerfringence. The charactirastics of MF including dispersion and nonlinearity are mainly depend on one parameter–diameter. Non-circular-symmetric (NCS) MFs with more freedom have been been proposed or demonstrated by exploiting different structures. With the fast development of MF fabrication technology, several approaches have been reported to fabricate NCS-MFs: such as adiabatically tapering conventional polarization maintaining (PM) fibers, rectangular-shaped fibers [5-7]. However, it presents great challenges to keep these asymmetric geometries during tapering, which is a serious concern.

In 2010, an alternative solution was proposed to realize a novel NCS-MF with a slot inside by post-processing a non-birefringent circular-symmetric MF (CS-MF) [8]. Similar with the slot waveguide in integrated optics [9] , a slot-MF can enhance and confine the light field in the nanometer-wide low index slot. Simulation results show that the birefringence reaches a high level of 4×10^{-2}, which is much higher than previous works [5-7]. By controlling the shape of the slot, light can be confined in the slot or not, then have higher or lower nonlinerity. A slot-MF is a hybrid of an MF and a slot waveguide including a larger evanescent field, high birefringence, compact size, flexible design, and simplified fabrication. It shows great potentials for miniature fiberized polarization manipulation devices, for example in-line MF wave plates. Its unique geometry can also enhance the refractive index sensitivity and offers opportunities for dispersion-related applications. Taking benefit of the special and flexible structure, it also can achieve ultra-wideband zero-dispersion. Finally, enhanced second harmonic generation (SHG) can be realized because of more surface area and stronger surface intensity. The fiberized slot waveguides (FSWs)-based devices may find more potential in sensor, laser and nonlinear signal processing with its unique geometry and extremely small size.

2. Geometry and Birefringence

A slot-MF can be fabricated by introducing a slot in an MF as illustrated in Figure 1 [8]. One or multiple straight or V-groove slots can be introduced in a MF. As a hybrid of a MF and a slot waveguide, this kind of device has many advantages including a larger evanescent field, compact size, flexible design, simplified fabrication, and low-loss interconnection to other optical fibers by maintaining the original size at their input–outputs. The slot can be formed with high accuracy by typical and popular micro-machining technologies, such as deep-UV lithography or focused ion beam milling [8]. These methods are time-saving and can achieve good sharpness of the slot edge.

In another way, it can be fabricated if we first drill a large slot in a thick MF and then re-taper it. The induced slot brings more freedom to optimize the linear and nonlinear charactrastics of the fiber. Here, the indices of the MF, medium in the slot and outside environment are denoted by n_f, n_s and n_c respectively. Other parameters used in this paper are presented in the cross section diagram (Figure 1). The fiber diameter $d = 2r$. All the calculation work is done at the wavelength of 1550 nm by using a full-vectorial finite-element analysis. All the simulations in this chapter are performed using the commercial finite element method (FEM) and matlab software. In this chapter, we assume $n_c = n_{air} \approx 1.00027$.

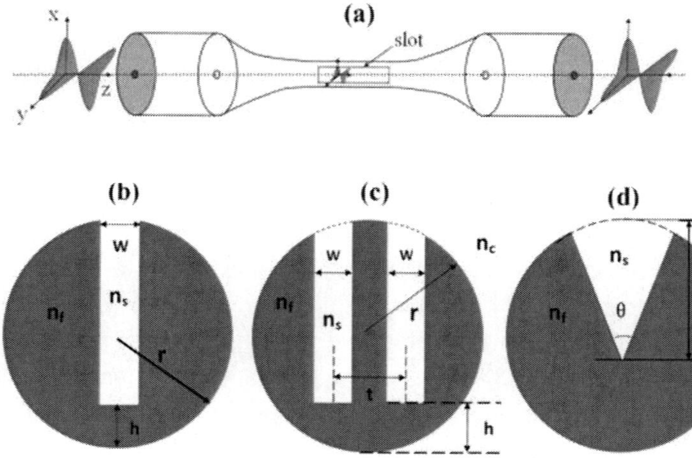

Figure 1. (a) Schematic of the slot-MF; Cross section of (b) a single-slot-MF, (c) a double-slot-MF and (d) a V-groove slot-MF. Here *W* denotes the width of the slot; θ denotes the V-groove angle, and *r* is the radius of the MF.

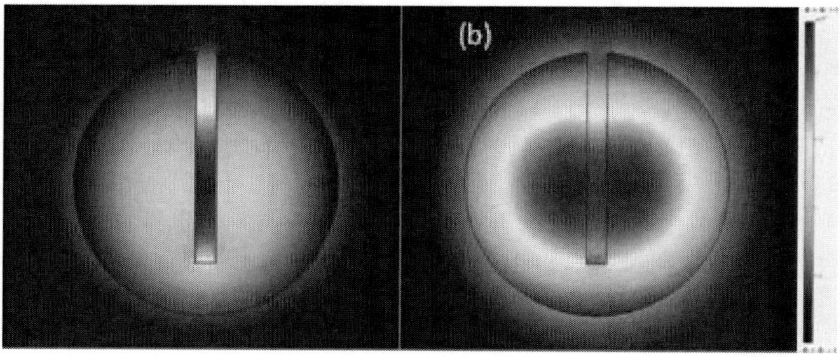

Figure 2. Mode field and power flow distribution of the fundamental HE_{11} modes with both x- and y-polarizations of a slot-MF ((a) & (b)) at d = 1000 nm, W = 80 nm and h = 200 nm in air and λ = 1550 nm.

Because of the asymmetrical geometry, slot-MF has high birefringence. We only consider the HE_{11} mode with two different polarizations in a single-slot-MF. The non-degeneracy of HE_{11} mode is known as phase birefringence B of the slot-MF, and is defined as: $B = n_{eff}^{y} - n_{eff}^{x}$, where n_{eff}^{x} and n_{eff}^{y} are effective refractive indices of HE_{11}^{x} and HE_{11}^{y} modes, respectively [8]. As shown in Figure 2, in order to satisfy the continuity of the normal component of electric flux density D (i.e., $D_{1n} = D_{2n}$), the corresponding electric field (E-field) must experience a large discontinuity with much higher amplitude in the slot side (the low index air) in the case of the HE_{11}^{x} mode. Due to the large-index-contrast interface of silica and air, much field power will be strongly confined in the slot region. which possesses a far lower refractive index (Figure 2 (a)) [8]. When the light wave is polarized in the y-direction, $D_{1n} = D_{2n}$ is naturally satisfied and light is mostly confined in the high index medium (Figure 2 (b)). Similar to planar slot waveguides, these modes (HE_{11}^{x} and HE_{11}^{y}) in slot-MFs are true-guided modes without confinement losses. The HE_{11}^{x} mode has very different properties and undergoes a different effective index compared with the HE_{11}^{y} mode [8].

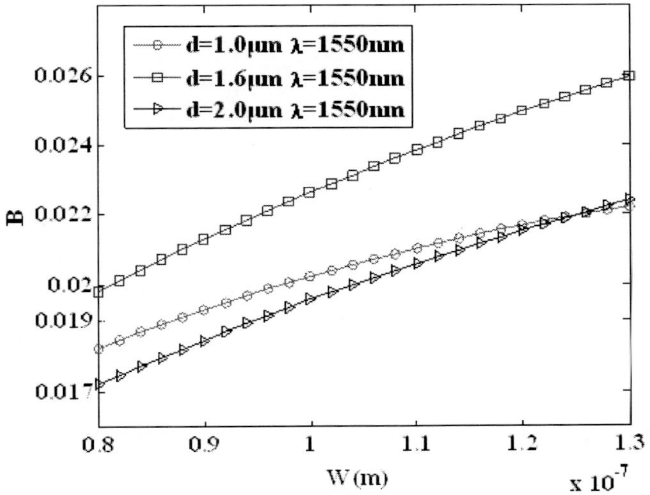

Figure 3. Birefringence of a slot-MF as a function of W.

As one of the most important parameters of slot-MFs, birefringence is calculated as shown in Figure 3 (λ = 1550 nm). Figure 3 shows the birefringence as a function of slot width and the radius of the MF. When the diameter and W are small, the mode field diameter is larger compared to that of the MF. So the evanescent field, which propagates outside the MF, contributes a big part of the total power flow. Thus, birefringence is relatively small. As the MF diameter and W increase, birefringence becomes larger and can reach a high level of > 2×10^{-2} at d = 2 μm with W > 100 nm. In fact, B can be as high as 0.04 at $d = 2$ μm and $W = 400$ nm. However, when the diameter of the MF is comparable to or bigger than the wavelength, birefringence decreases. In this case, the light fields are well confined in two D-shaped silica rods, resulting in a low birefringence [8].

Considering the fabrication feasibility of the slot-MF, we also numerically study the influence of slot shift on birefringence. It shows that little change of birefringence is observed with slot shift smaller than 100 nm compared to that of the slot located at the center of the MF. Thus, we can easily fabricate slot-MFs with birefringence as large as 10^{-2} using MFs of different radius.

Finally, $h = 0$ is an extreme case and it is possible but maybe a little difficult to fabricate and handle. For practical applications, it is more preferred at $0 < h < \mathrm{r}$.

3. Dispersion

The Group Velocity Dispersion (GVD) or chromatic dispersion is a primary cause of concern when dealing with short duration pulses at high bit rate single-mode WDM systems because light from a typical optical source contains different wavelength components which travel at different speeds. Low-chromatic-dispersion is also very important for linear applications to maintain low distortion of analog or digital signals and high-bandwidth data signals transmission. There are two main results of material dispersion and waveguide dispersion. Controlling and tuning the dispersion is very important in all fiber linear and nonlinear applications. In a standard single mode fiber (SMF), dispersion causes pulses to spread and it has to be compensated using dispersion compensating fibers or devices, which are designed to have large negative dispersion. In nonlinear frequency conversion, the phase matching is decided by the dispersion profile and supercontinuum generation typically takes place around the zero point of the GVD. Another key challenge for dispersion management in optical fibers and integrated waveguides is to realize a very flat and low dispersion over a wide wavelength range. Wideband near-zero dispersion has been therotically proposed or experimentally demonstrated in optical photonic crystal fibers (PCFs), which are made from undoped fused silica with an artificial array of air holes running along its length allowing flexible tailoring of the dispersion curves. A number of designs have been reported by manipulating the geometry of the PCF holes, for example, designing a PCF containing elliptical air-holes [10], selectively filling some of the air-holes with liquid[11], varying the diameters of the air-hole rings along the fiber radius [12], and using a PCF with defected-core [13].

The total dispersion of a fiber can be calculated by [14, 15].

$$D(\lambda) = -\frac{\lambda}{c}\frac{d^2 n_{eff}}{d\lambda^2} \tag{1}$$

where c is the velocity of light in vacuum, λ is the wavelength and n_{eff} is the effective refractive. The material dispersion given by Sellmeier's formula [16] has been taken into consideration in the calculation.

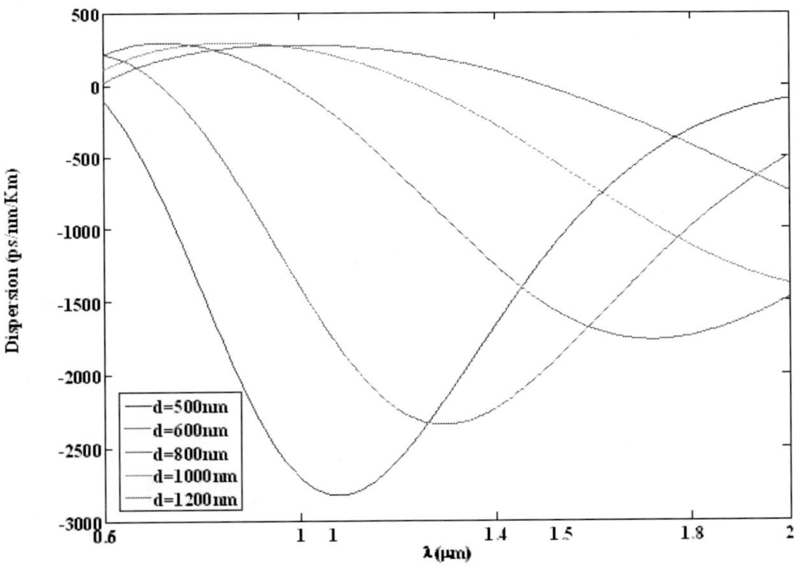

Figure 4. The calculated dispersion profile of a CS-MF with different diameters. The dispersion value changes quickly with the wavelength and it is very difficult to achieve a small dispersion over a wide wavelength range only by tuning the MF diameter.

For conventional standard SMF, the magnitude of the dispersion parameter is of the order of 10…20 ps/(nm km). If we want to modify the dispersion the main method is to change the diameter by tapering. Figure 4 shows the dependence of CS-MF dispersion on the wavelength for five different diameters. The dispersion of the tapered fiber depends strongly on the diameter of the taper waist, the dispersion values are quite large in the normal-dispersion regime, ranging in the order of thousands of ps/km/nm, and on the other hand, it can be reduced or even made anomalous at visible and near-infrared wavelengths. The zero-dispersion wavelength (ZDW) can be shifted from 1270 nm to wavelengths below 700 nm. At the same time another ZDW is appearing at the near-infrared wavelengths. Figure 5 shows the dependence of the CS-MF ZDWs on *d* for wavelengths in the range of 0.6 - 2 μm. There are two zero dispersion points when *d* is between 1300-1600 nm and the wavelength is between 0.6 - 2 μm. The large difference between the refractive indices of the fiber material and air limits the position and the extent of spectral broadening. However, it is a challenging task to flatten the dispersion curves even in air-cladding sub-wavelength tapered SMFs with symmetric and circular cross-sections [1, 17, 18]. Furthermore there are strong nonlinear optical properties due to the tight light confinement in the solid MFs [15].

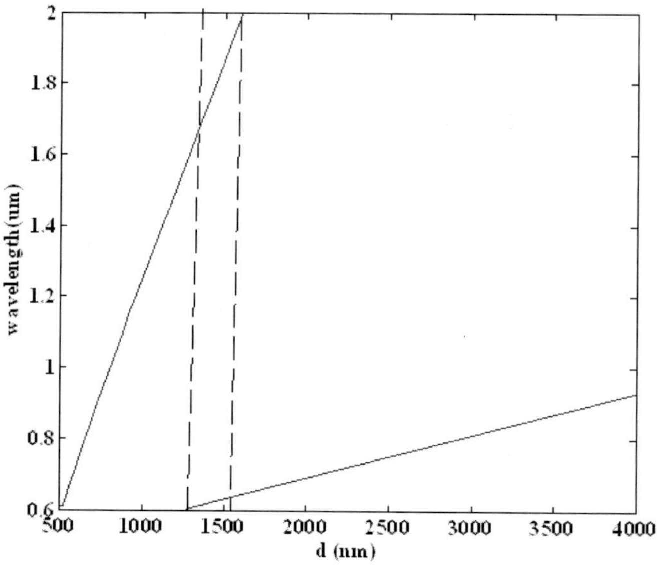

Figure 5. The ZDW of CS-MF in the range of 0.6 - 2 μm as a function of *d*.

For this reason, one expects to be able to control, at least to some extent, the dispersion properties of guided modes by breaking the symmetrical geometry of the CS-MF. The slot-MF, like the planar slot-waveguide, has additional degrees of freedom in design based on the slot size as compared to the CS-MF, and it is possible to realize ultra-flattened low dispersion over a wide wavelength range [15].

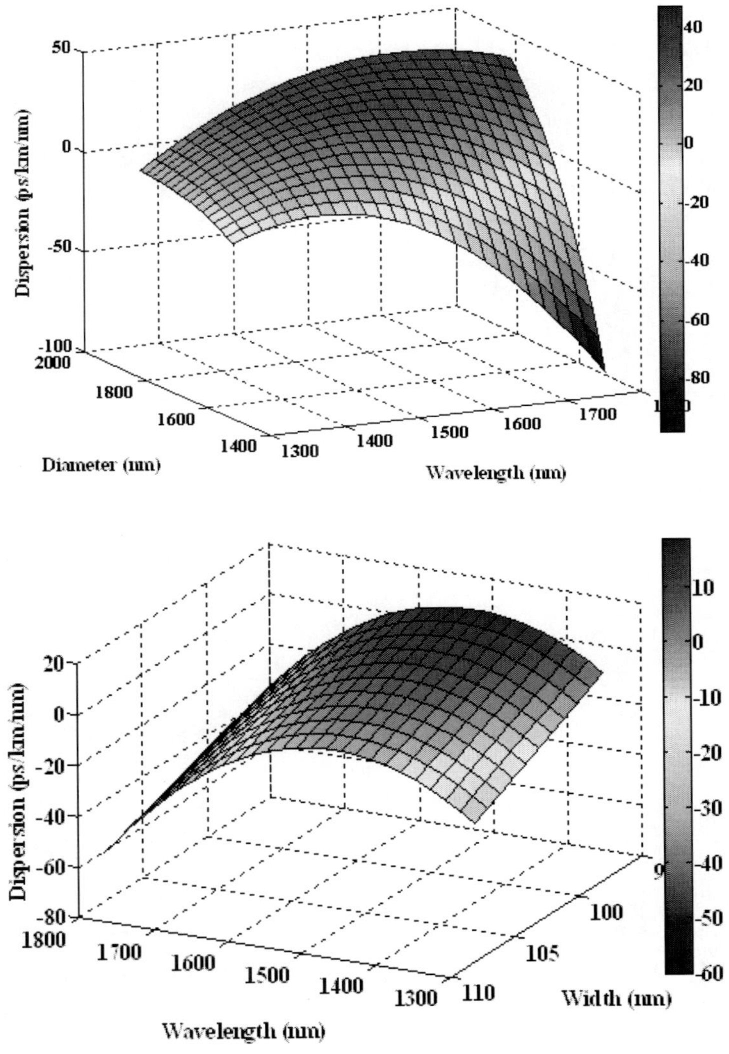

Figure 6. Dispersion profile evolution in a single-slot-MF by tuning (a) the fiber diameter at $W = 100$ nm and (b) the slot width at $d = 1.6$ μm.

For a single-slot-MF, Figure 6 (a) and (b) shows the dependence of the dispersion profiles on the fiber diameter d and the slot width W, respectively. As d increases from 1520 to 1920 nm, the dispersion value increases and the two ZDWs appear when $d > 1540$ nm and shift away from each other. As W increases from 98 to 108 nm, the dispersion value increases and the two ZDWs shift toward each other and disappear when $W > 105$ nm [15]. The flattened near-zero dispersion curve appears around W = 100 nm and d = 1.6μm, respectively. Figure 7 shows two selected flattened nearly-zero dispersion curves. One has low dispersion of ±10 ps/(nm·km) over bandwidth in excess of 340 nm from a wavelength of 1320 nm to 1660 nm. Another one has low dispersion value of -5 to 10 ps/(nm·km) over a 340 nm wavelength range, from 1300 nm to 1640 nm [15].

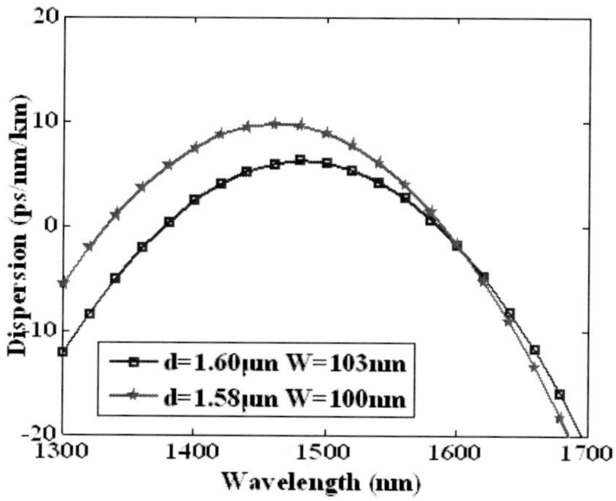

Figure 7. Two selected flattened near-zero dispersion profiles with optimized parameters.

We also can do some simulation results on double-slot-MF, and the dispersion values drop as the width W increases or diameter d decreases, which is similar to the conclusion we draw from the single-slot-MF. The pitch t has a similar influence on the dispersion profile with d, however, the ZWDs shift in the opposite directions. Comparing the results of the single-slot and double-slot-MFs, there is no big difference except for the red-shift of the nearly-zero dispersion region in the double-slot-MFs [15].

4. Third-Order Nonlinearity

The third-order nonlinearity of silica-based fiber is very small, however, light can be confined to a small cross-section over long lengths, which makes the power density high and the nonlinear effects quite important in the design of modern fiber optical systems. Optical nonlinearities can significantly degrade the information capacity and limit the channel power in high-speed optical communications. On the other hand, considerable effort has been directed toward the applications of nonlinear phenomena such as self-phase modulation (SPM), cross-phase modulation (XPM), stimulated Raman scattering (SRS), and four-wave mixing. All of these nonlinear effects have being explored to realize a number of optical functions including wavelength conversion, optical switching, slow or fast light. Generally, the nonlinear coefficient in un-uniform waveguides with nanoscale structures can be defined as [14].

$$\gamma = k(\frac{\varepsilon_0}{\mu_0})\frac{\int n^2(x,y)n_2(x,y)[2|e|^4 + |e^2|^2]dA}{3\left|\int (e \times h^*).\hat{z}dA\right|^2} \tag{2}$$

where n and n_2 are the spatial distribution of the refraction index and nonlinear-index coefficients, e and h stand for the spatial distribution of electric and magnetic fields. For different applications, reduced or enhanced nonlinearity must be considered by optimizing the waveguide geometry and material components.

CS-MF displays strong nonlinearity because light can be confined in a much smaller nonlinear region allowing the ready observation of nonlinear interactions at relatively modest power levels. As shown in Figure 8, the fraction of power in the fiber can be as high as 92 % at $d = 1$ μm, which means that the power intensity is nearly 10,000 times the standard 125 μm SMF. Figure 9 depicts the nonlinear coefficient of a CS-MF with different diameters as a function of wavelength. It is obvious that with smaller diameters we can obtain larger nonlinear coefficient values because the field intensity of the fiber becomes larger as the diameter decreases. The typical nonlinearity coefficient of a CS-MF at $\lambda = 1550$ nm and $d = 1400$ nm is ~1 $W^{-1}m^{-1}$, much higher than that of a standard telecom SMF which is ~1 W^{-1} km^{-1}. CS-MFs made from highly nonlinear materials such as lead silicate, bismuth silicate, and chalcogenide glasses can be used to produce higher nonlinearity and can be used for a wide range of nonlinear applications, including supercontinuum generation and optical switching.

In a slot-MF, some of the power is confined in the slot and the material in the slot is important for the nonlinearity. Figure 10 shows the fractional power as a function of W: more than 30% of the power is in the slot, nearly half of the power in the fiber.

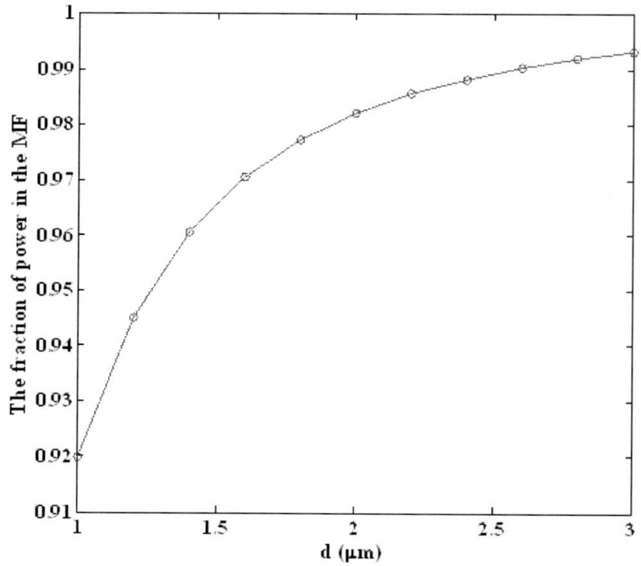

Figure 8. The fraction of power in the fiber as a function of the MF diameter *d*. Here the environment is air, and the wavelength is 1550 nm.

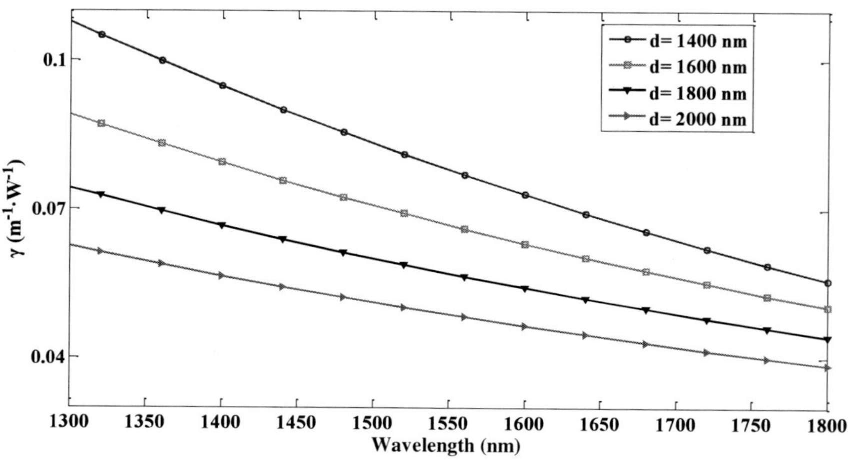

Figure 9. Nonlinear coefficient profiles of a CS-MF with different diameters.

For an air-slot-MF, since part of the light is confined within the low refractive index, the linear air region, the nonlinear coefficient (γ) of the waveguide can be deduced.

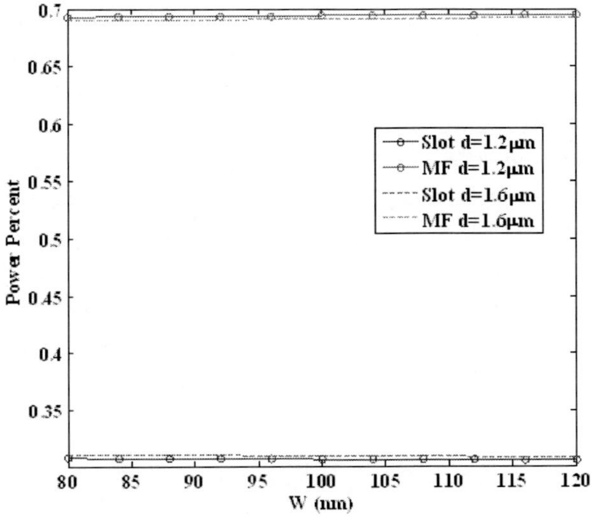

Figure 10. The fraction of power in the slot and MF as a function of *W*. Here the environment is air, and the wavelength is 1550 nm.

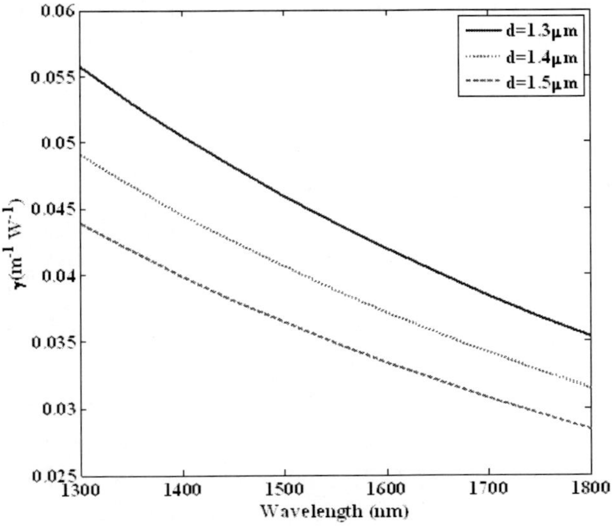

Figure 11. Nonlinear coefficient profiles of a single slot-MF with different diameter. Here $W = 90$ nm and $h = 0.2\ d$.

In Figure 9, the nonlinear coefficient profile at different *d* is plotted. It shows that the nonlinear coefficient value increases with the decrease of fiber diameter because more light is confined in the near-zero nonlinear-index coefficient air-slot region and makes little contribution to the nonlinear coefficient. It can be seen that the nonlinear coefficient value can drop to ~ 40% for a single-slot-MF, compared to a CS-MF with the same diameter.

On the other hand, the nonlinear coefficient can be tuned by changing the geometry parameter and material of the slot. If we choose high nonlinear materials in the slot, it is possible for the nonlinear coefficient to be enlarged. Of course, special fiber materials with a relatively higher index must be used to keep the lower-index slot area.

5. Surface Nonlinearity

Second-harmonic generation (SHG) in a non-central-symmetry crystal structure can be easily observed and depends mainly on two factors: the second-order nonlinear susceptibility (χ^2) and phase matching of the fundamental and the second-harmonic waves. However, unlike its related quartz crystalline phase, fused silica fiber is amorphous with a macroscopic inversion symmetry and it is supposed that all of the second-order dipole nonlinear coefficients should be zero. Second-order nonlinearities should not appear in silica optical fibers. Despite this, some previous experiments showed that SHG could be observed. The discovery by Österberg and Margulisin 1986 of SHG in an optical fiber with power-conversion efficiency as high as ~3%, has generated considerable interest in the physics and applications of this unexpected phenomenon [19]. In 1987 they reported efficient SHG in phosphor-doped glass fibers with higher peak conversion efficiencies > 5% in fibers less than a meter long after the harmonic had built up over many hours, using input intensities of 100 – 200 GW/cm^2 [20]. Then, more research about the SHG mechanism and experiments were reported by the end of the 1980s [21-23]. More recently, several techniques including thermal poling, corona poling, and electron implantation have been employed to introduce an effective χ^2 [24, 25]. However, the exact nature of SHG in optical fibers is still unclear. Initially, this phenomenon could not be reasonably explained by core-cladding interface or bulk multipole moment contributions to the second-order nonlinearity [26]. A subsequent study showed that the nonlinearity caused by the formation of a second-order susceptibility ($\chi^{(2)}$) grating through multiphoton processes involving both the pump and SHG light agreed with the experimental results [27]. The $\chi^{(2)}$ grating introduces the second-order polarization and compensates the phase mismatch. Because of a self-saturation effect by the interference of the SHG light with the $\chi^{(2)}$ grating, it is impossible to have the SHG conversion efficiency beyond the level of a few percent [27].

A recent experiment showed phase-matched SHG at 532 nm in a low-order mode of a CS-MF. The multiphoton processes that induced $\chi^{(2)}$ grating are limited in silica MFs for the weak photosensitivity. In fact, the sub wavelength size of MFs enhance the power density and increase the interface and bulk multipole moment contributions to $\chi^{(2)}$, so that surface nonlinearity of core-cladding interface and nonlinearity of the bulk multipole become the major mechanism for SHG. Furthermore, large core-cladding index contrast makes it possible to achieve SHG phase matching in a low-order mode with sufficient intensity at the surface [28]. In order to enhance the surface SHG effect in MFs, more surfaces with strong surface field are preferred. By inducing a slot structure in the MF, a great power density around the slot can help to enhance the surface nonlinearity. Here we investigate the surface SHG by phase matching between fundamental and low-order mode in slot-MFs, and compare the results with that of CS-MFs. CS-MFs are studied by analytical methods, and slot-MFs will be studied numerically using the FEM. It can be seen that higher SHG conversion will theoretically be achieved in slot-MFs. The maximum conversion efficiency in our calculations is about 24 times larger than that of CS-MFs.

Using small-signal approximation (pump depletion is negligible), the SHG process can be described by the amplitudes coupling equation [29, 30]:

$$\frac{dA_2}{dz} - i\rho_2 A_1^2 \exp(i\Delta\beta z) = 0 \tag{3}$$

where A_1 and A_2 are the field amplitudes of the pump and SHG signals, respectively; $\Delta\beta=2\beta_1-\beta_2$ is the phase mismatch between the fundamental and second-harmonic waves; ρ_2 is the overlap integral [29, 30]:

$$\rho_2 = \frac{\omega_2}{4N_1\sqrt{N_2}} \int \mathbf{e}_2^* \cdot \mathbf{P}^{(2)} dS \tag{4}$$

where ω_2 is the second-harmonic angular frequency. The function will be integrated on the cross-sectional region of the MF, and dS is the area element. All the field components are normalized by the normalizing factors [29, 30]:

$$N_j = \frac{1}{2}\int |(\mathbf{e}_j^* \times \mathbf{h}_j)\cdot\hat{\mathbf{z}}|\, dS \ (\mathrm{j}=1,2) \tag{5}$$

The fields of the guided modes can be written as [29, 30]:

$$\mathbf{E}(\mathbf{r},\omega_j) = A_j(\omega_j)\mathbf{e}_j(\mathbf{r},\omega_j)\exp(i(\beta_j z-\omega_j t)) \tag{6}$$

$$\mathbf{H}(\mathbf{r},\omega_j) = A_j(\omega_j)\mathbf{h}_j(\mathbf{r},\omega_j)\exp(i(\beta_j z-\omega_j t)) \tag{7}$$

$P^{(2)}$ is the second-order nonlinear polarization. For pure-silica MFs in air cladding, it originates from the contributions of silica-air interface and bulk multipole moments. The bulk contributions are expressed as [12, 13]:

$$\mathbf{P}_b^{(2)}(\mathbf{r}) = \varepsilon_0\gamma\nabla(\mathbf{E}_1\cdot\mathbf{E}_1) + \varepsilon_0\delta(\mathbf{E}_1\cdot\nabla)\mathbf{E}_1 \tag{8}$$

A third term proportional to $\mathbf{E}_1(\nabla\cdot\mathbf{E}_1)$ can be included in the surface term [2]. The bulk contributions shown in eq. (8) can be ignored, since the results for the silica MFs indicated them to be of minor importance [29, 30]. So we just take surface contributions into account, and $\mathbf{P}^{(2)}$ can be written as:

$$\mathbf{P}^{(2)} \approx \mathbf{P}_s^{(2)}(\mathbf{r}) = \delta(\mathbf{r}-\mathbf{S})[\mathbf{P}_{\perp}^{(2s)} + \mathbf{P}_{\perp\parallel}^{(2s)} + \mathbf{P}_{\parallel}^{(2s)}] \tag{9}$$

$$\begin{cases} \mathbf{P}_{\perp}^{(2s)} = \varepsilon_0\chi_{\perp}^{(2s)}\mathbf{e}_{1\perp}^2\hat{\mathbf{r}}_{\perp} \\ \mathbf{P}_{\perp\parallel}^{(2s)} = \varepsilon_0\chi_{\perp\parallel}^{(2s)}\mathbf{e}_{1\parallel}^2\hat{\mathbf{r}}_{\perp} \\ \mathbf{P}_{\parallel}^{(2s)} = 2\varepsilon_0\chi_{\parallel}^{(2s)}e_{1\perp}\mathbf{e}_{1\parallel} \end{cases} \tag{10}$$

where $\hat{\mathbf{r}}_{\perp}$ is the unit vector normal to the interface. The three terms of surface second-order susceptibility can be measured by experiments: $\chi_{\perp}^{(2s)} = 6.3\times10^3\,\mathrm{pm^2/V}$, $\chi_{\perp\parallel}^{(2s)} = 7.7\times10^2\,\mathrm{pm^2/V}$, $\chi_{\parallel}^{(2s)} = 7.9\times10^2\,\mathrm{pm^2/V}$ [29, 30].

For CS-MF, the fields of the guided modes are available by solving the Maxwell equations. In our calculation, the pump signal and SHG modes are assumed to be the HE_{11} mode and the HE_{21} mode, respectively. The phase matching condition $\Delta\beta=2\beta_1-\beta_2=0$ is satisfied by material dispersion and mode dispersion of the fiber. The material dispersion of silica glass can be described by the Sellmeier polynomial [30].

The SHG signal will only efficiently generate in the phase-matched mode with $\Delta\beta=2\beta_1-\beta_2=0$. In order to realize phase matching in different SHG frequencies, we modify the mode dispersion by changing the fiber diameter.

In the small-signal limit with perfect phase matching, the power-conversion efficiency is given as [26]:

$$\frac{P_2}{P_1} = (\rho_2 z)^2 P_1 \tag{11}$$

where P_1 is the pump power and z is the interaction length (generally the waist length of MF or the slot length). This is not an accurate description of SHG dynamics, but it is sufficient to estimate the SHG conversion efficiencies for comparison between CS-MFs and slot-MFs.

According to eq. (11), SHG conversion efficiency is proportional to the square of the overlap integral ρ_2 with all other conditions being equal. Thus the absolute value of ρ_2 determines the SHG conversion capability. From eq. (3) – (11) and analytic solutions of mode fields, we can calculate the $|\rho_2|$ of CS-MFs.

For slot-MFs, $|\rho_2|$ can be calculated with the numerical modeling method. Here mode fields and propagation constants are determined using the FEM. For slot-MFs, the pump signal is assumed to propagate in the HE_{11}-like fundamental mode, while the SHG signal is assumed to generate in the HE_{21}-like mode (the 5th-order mode). The $|\rho_2|$ of slot-MFs is determined by eq. (3) – (10) and numerical solutions of mode fields and propagation constants.

Figure 12 shows the calculated relationship between the fiber diameter and the phase-matched SHG wavelength for CS-MF and slot-MF (h = 0.5d, W = 0.05d). As the slot number increases, phase-matched λ_{SHG} for different structures at the same diameter decreases. It results from the modulation of the waveguide dispersion by slot structure parameters. The slot structure enlarges the waveguide dispersion by more evanescent field propagating outside the fiber, making β_1 of the pump wave in the fundamental mode and β_2 of the second-harmonic wave in a high-order mode matched at a shorter wavelength. The modulation can be enhanced by adding more slots in the fiber [30].

Figure 13 shows the power flow distribution of CS-MF and slot-MF in the HE_{21} or HE_{21}-like mode for the corresponding phase-matched λ_{SHG}. In the figure, we can see that the slot increases the surface area and a fraction of the light field is confined in the slot structure and enlarges the surface power density [30].

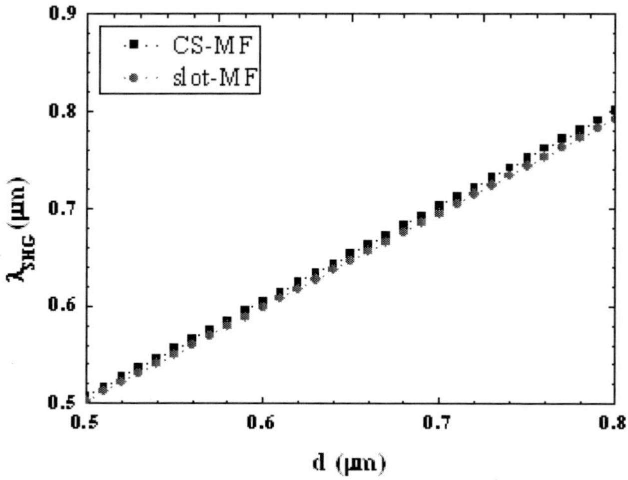

Figure 12. The relation between phase-matched SHG wavelength λ_{SHG} and the fiber diameter d in CS-MF (black line) and slot-MF (red line).

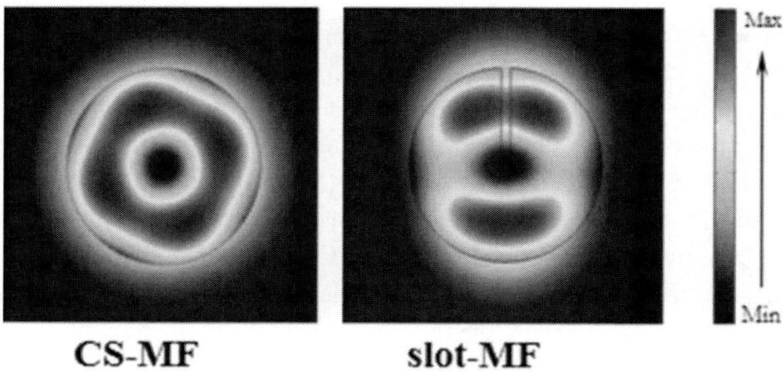

Figure 13. Power flow distribution of CS-MF (left) and slot-MF (right) in HE_{21} or HE_{21}-like mode.

In Figure 14, the absolute value of ρ_2 for CS-MF, slot-MF1 ($h = 0.5d$, $W = 0.05d$), slot-MF2 ($h = 0$, $W = 0.05d$), is plotted versus the phase-matched λ_{SHG}. Obviously, the surface area scales up with the height of the slots. With concentrated power distribution and a large surface area in the slot, $|\rho_2|$

roughly scales with $(\lambda_{SHG})^{-3}$ in all these structures. $|\rho_2|$ of slot-MFs is significantly larger than that of CS-MF, and slot-MF2 has the maximum $|\rho_2|$ (about 5 times of that in CS-MF). This can be explained by increasing the surface area and the power density at the surface. More surface area and higher surface power density contribute to stronger surface nonlinearity.

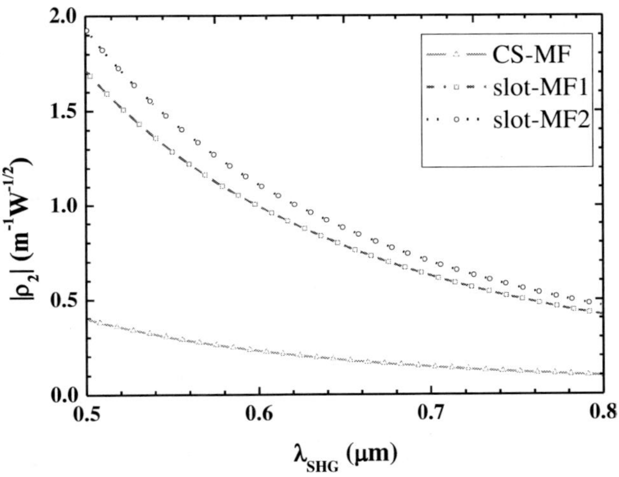

Figure 14a. Absolute value of overlap integral $|\rho_2|$ versus λ_{SHG} for CS-MF (orange line), slot-MF1 (green line) and slot-MF2 (black line).

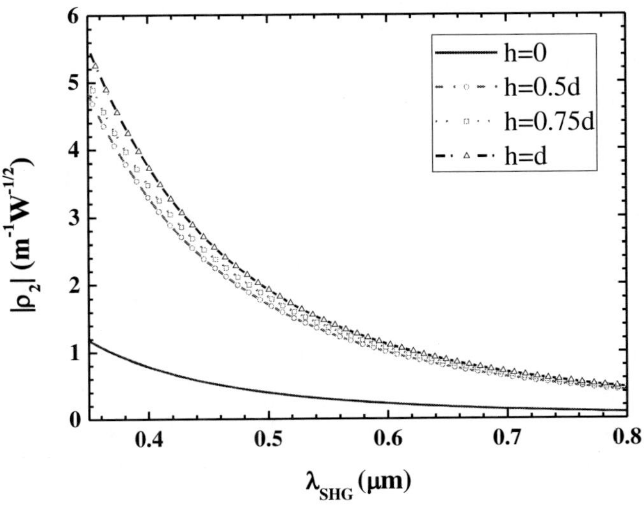

Figure 14b. $|\rho_2|$-λ_{SHG} relation for slot-MFs with different slot heights when $\boldsymbol{W} = 0.05d$.

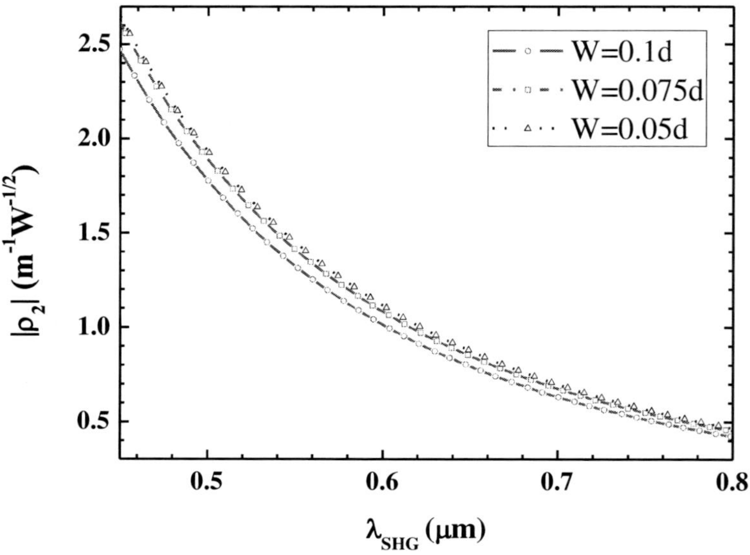

Figure 15. $|\rho_2|$-λ_{SHG} relation for slot-MFs with different slot widths when $h = 0$.

Figure 14 and 15 show the relation between $|\rho_2|$ and λ_{SHG} depending on the modulation of W and h in the slot-MF. $|\rho_2|$ decreases as W increases or h increases. A smaller h means more surface area and a stronger surface effect. A wider slot dispersed the power flow distribution, which leads to a lower light intensity at the surface and then the reduction of $|\rho_2|$. Further calculation on the double-slot-MF shows that the distance between slots has little influence on the overlap integral [30].

According to eq. (11), the absolute value of ρ_2 determines the SHG conversion efficiency. From the simulation results, it proves that the SHG conversion efficiency achieved in slot-MFs is significantly higher than in CS-MFs. For slot-MF2 of slot length $z = 1$ mm and pumping peak power $P_1 = 1$ kW with a 1550 nm femtosecond laser source, conversion efficiency is calculated to be ~ 0.027%, while a CS-MF with the same parameters only has a conversion efficiency ~ 0.0011%. Further improvement of the SHG efficiency can be realized by optimizing the structural parameters and utilizing other mechanism of second-order nonlinearity. For example, strain in the material breaks the symmetry of the structure, introducing sizeable second-order nonlinearity into the waveguide, so that a stressing over layer can be deposited on the fiber to enhance the nonlinearity [17].

CONCLUSION

In this chapter, we presented the so-called slot-MF with a nanoscale slot inside the MF. Its significant linear and nonlinear characteristics are numerically calculated and discussed. With its unique geometry, the slot-MF shows several attractive performances including high birefringence, large evanescence field, flexible dispersion, nonlinear profiles and a strong surface effect. In particular, its dispersion profile can be optimized when flattened and near zero in a wide wavelength range, and enhanced SHG can be realized because of the larger surface area and power density at the surface in slot-MFs. These advantages present great potential in sensing, lasing and communication applications.

ACKNOWLEDGMENTS

This work is supported by National 973 program under contract No. 2012CB921803 and 2011CBA00205, NSFC program No. 11074117 and the National Science Fund for Excellent Young Scientists Fund (61322503) and National Science Fund for Distinguished Young Scholars (61225026). The authors also acknowledge the support from PAPD and the Fundamental Research Funds for the Central Universities.

REFERENCES

[1] Tong, L. M.; Gattass, R. R.; Ashcom, J. B.; He, S. L.; Lou, J. Y.; Shen, M. Y.; Maxwell, I.; Mazur, E., Subwavelength-diameter silica wires for low-loss optical wave guiding. *Nature* 2003, 426, (6968), 816-819.

[2] Bilodeau, F.; Hill, K. O.; Johnson, D. C.; Faucher, S., Compact, Low-Loss, Fused Biconical Taper Couplers - Overcoupled Operation and Antisymmetric Supermode Cutoff. *Optics Letters* 1987, 12, (8), 634-636.

[3] Bilodeau, F.; Hill, K. O.; Faucher, S.; Johnson, D. C., Low-Loss Highly Overcoupled Fused Couplers - Fabrication and Sensitivity to External-Pressure. *Journal of Lightwave Technology* 1988, 6, (10), 1476-1482.

[4] Brambilla, G.; Xu, F.; Horak, P.; Jung, Y.; Koizumi, F.; Sessions, N. P.; Koukharenko, E.; Feng, X.; Murugan, G. S.; Wilkinson, J. S.;

Richardson, D. J., Optical fiber nanowires and microwires: fabrication and applications. *Advances in Optics and Photonics* 2009, 1, (1), 107-161.

[5] Jung, Y.; et al., Polarization-maintaining optical microfiber. *Opt. Lett.* 2010, 35, (12), 2034-2036.

[6] Jung, Y.; Brambilla, G.; Oh, K.; Richardson, D. J., Highly birefringent silica microfiber. *Opt. Lett.* 2010, 35, (3), 378-380.

[7] Xuan, H.; Ju, J.; Jin, W., Highly birefringent optical microfibers. *Opt. Express* 2010, 18, (4), 3828-3839.

[8] Jun-long, K.; Fei, X.; Lu, Y.-q., Highly Birefringent Slot-Microfiber. *Photonics Technology Letters, IEEE* 2011, 23, (15), 1034-1036.

[9] Almeida, V. R.; Xu, Q. F.; Barrios, C. A.; Lipson, M., Guiding and confining light in void nanostructure. *Opt. Lett.* 2004, 29, (11), 1209-1211.

[10] Hai, N. H.; Namihira, Y.; Kaijage, S. F.; Kinjo, T.; Begum, F.; Razzak, S. M. A.; Zou, N., A unique approach in ultra-flattened dispersion photonic crystal fibers containing elliptical air-holes. *Opt Rev* 2008, 15, (2), 91-96.

[11] Gundu, K. M.; Kolesik, M.; Moloney, J. V.; Lee, K. S., Ultra-flattened-dispersion selectively liquid-filled photonic crystal fibers. *Opt Express* 2006, 14, (15), 6870-6878.

[12] Saitoh, K.; Koshiba, M., Highly nonlinear dispersion-flattened photonic crystal fibers for supercontinuum generation in a telecommunication window. *Opt Express* 2004, 12, (10), 2027-2032.

[13] Saitoh, K.; Florous, N.; Koshiba, M., Ultra-flattened chromatic dispersion controllability using a defected-core photonic crystal fiber with low confinement losses. *Opt Express* 2005, 13, (21), 8365-8371.

[14] Shahraam, A. V.; Monro, T. M., A full vectorial model for pulse propagation in emerging waveguides with subwavelength structures part I: Kerr nonlinearity. *Opt Express* 2009, 17, (4), 2298-2318.

[15] Guo, W.; Kou, J.-l.; Xu, F.; Lu, Y.-q., Ultra-flattened and low dispersion in engineered microfibers with highly efficient nonlinearity reduction. *Optics Express* 2011, 19, (16), 15229-15235.

[16] Okamoto, K., *Fundamentals of optical waveguides*. 2nd ed.; Elsevier: Amsterdam; Boston, 2006; p xvi, 561 p.

[17] Brambilla, G.; Finazzi, V.; Richardson, D., Ultra-low-loss optical fiber nanotapers. *Opt. Express* 2004, 12, (10), 2258-2263.

[18] Tong, L. M.; Lou, J. Y.; Mazur, E., Single-mode guiding properties of subwavelength-diameter silica and silicon wire waveguides. *Opt Express* 2004, 12, (6), 1025-1035.

[19] Österberg, U.; Margulis, W., Dye laser pumped by Nd:YAG laser pulses frequency doubled in a glass optical fiber. *Opt. Lett.* 1986, 11, (8), 516-518.

[20] Österberg, U.; Margulis, W., Experimental studies on efficient frequency doubling in glass optical fibers. *Opt. Lett.* 1987, 12, (1), 57-59.

[21] Valk, B.; Kim, E. M.; Salour, M. M., Second harmonic generation in Ge-doped fibers with a mode-locked Kr+ laser. *Applied Physics Letters* 1987, 51, (10), 722-724.

[22] Farries, M. C.; Russell, P. S. J.; Fermann, M. E.; Payne, D. N., Second-harmonic generation in an optical fibre by self-written χ₍₂₎ grating. In *Electronics Letters*, Institution of Engineering and Technology: 1987; Vol. 23, pp 322-324.

[23] Gabriagues, J. M.; Février, H., Analysis of frequency-doubling processes in optical fibers using Raman spectroscopy. *Opt. Lett.* 1987, 12, (9), 720-722.

[24] Myers, R. A.; Mukherjee, N.; Brueck, S. R. J., Large second-order nonlinearity in poled fused silica. *Opt. Lett.* 1991, 16, (22), 1732-1734.

[25] Canagasabey, A.; Corbari, C.; Gladyshev, A. V.; Liegeois, F.; Guillemet, S.; Hernandez, Y.; Yashkov, M. V.; Kosolapov, A.; Dianov, E. M.; Ibsen, M.; Kazansky, P. G., High-average-power second-harmonic generation from periodically poled silica fibers. *Opt. Lett.* 2009, 34, (16), 2483-2485.

[26] Terhune, R. W.; Weinberger, D. A., Second-harmonic generation in fibers. *J. Opt. Soc. Am. B* 1987, 4, (5), 661-674.

[27] Anderson, D. Z.; Mizrahi, V.; Sipe, J. E., Model for second-harmonic generation in glass optical fibers based on asymmetric photoelectron emission from defect sites. *Opt. Lett.* 1991, 16, (11), 796-798.

[28] Cheng, T.; Gao, W.; Kawashima, H.; Deng, D.; Liao, M.; Matsumoto, M.; Misumi, T.; Suzuki, T.; Ohishi, Y., Widely tunable second-harmonic generation in a chalcogenide– tellurite hybrid optical fiber. *Opt. Lett.* 2014, 39, (7), 2145-2147.

[29] Lægsgaard, J., Theory of surface second-harmonic generation in silica nanowires. *J. Opt. Soc. Am. B* 2010, 27, (7), 1317-1324.

[30] Luo, W.; Guo, W.; Xu, F.; Lu, Y.-q., Efficient surface second-harmonic generation in slot micro/nano-fibers. *Optics Express* 2013, 21, (9), 11554-11561.

In: Nonlinear Optics
Editor: Fred Wilkins

ISBN: 978-1-63321-926-7

Chapter 5

GENERATION OF 2ND HARMONIC LIGHT FROM NONCENTROSYMMETRIC MATERIAL

Kitsakorn Locharoenrat*
Department of Physics, Faculty of Science,
King Mongkut's Institute of Technology Ladkrabang,
Bangkok, Thailand

ABSTRACT

Over the past few years significant progress has been made in investigations into the synthesis of nanowires of various materials and their applications to nanoscale electronics, optical devices and sensing systems; nevertheless, a major hurdle facing the developers of the integrated nanowire systems is organizing, manipulating, or controlling the placement of large numbers of nanowires. These problems can be solved by using a shadow deposition method for fabrication of polycrystalline metallic nanowires. This technique uses topographical features of ridge-to-valley to expose specific areas of the alkali halide substrates in order to grow the nanowire arrays. Nanowires of large areas show a preference for formation along exposed steps in the atomic structure of the substrate. The minimum widths of the obtained nanowires are less than 40 nm.

* Tel.: +66-2329-8000. Fax: +66-2329-8412. E-mail: klkitsak@kmitl.ac.th (Locharoenrat K.).

The strongly geometrical shape, high aspect ratios, and nanoscale cross-section of metallic nanowires can be expected to affect optical properties through confinement effects. Optical properties of the obtained metallic nanowires are then investigated by studies of linear reflectivity spectra, linear absorption spectra, and second harmonic generation spectra. Theoretical calculation by Maxwell-Garnett model and finite-difference time-domain method are also represented for predicting the reflectivity, absorption and the electric field distributions close to the wires. With experimental and theoretical studies, it is discovered that there are differences of optical properties among these metallic nanowires. These will be one of the most important issues when considering the types of materials used in current applications and development of new applications.

I. Introduction

There are many potential applications for which metallic nanowires may become important: for electronic nanodevices and for metallic interconnects of quantum nanodevices. These applications in any devices require the large areas with high throughput of the metallic nanowires. A shadow deposition method is considered as one of the promising procedures to fulfill this requirement. In contrast to a lithographic method [1], this self-organized pattern formation is also completed without breaking vacuum [2].

The primary role of the small size of the metallic nanowires shows a wealth of optical phenomena. The external light photons can couple with the electrostatic field of plasmons generating the strong absorption band [3]. With the development of pulsed laser technology, nonlinear optical techniques including second harmonic generation (SHG) are also used to understand the anisotropic nature of these nanomaterials. A great advantage of these studies over the linear optical techniques is its intrinsic surface sensitivity due to the fact that the SH signal originates from a thin interfacial region [4-6]. Owing to this surface sensitivity, the resonant coupling of surface plasmons in the metallic nanowires with SHG may result in an enhancement of the electromagnetic field. The linear dielectric response of the composites of the metallic nanowires in the dielectric medium may lead to an enhancement of second order optical nonlinearity. These factors motivate the need for further studies towards the elucidation of the microscopic origin of the nonlinearity of metallic nanostructures. In this chapter, three types of metallic nanowires are fabricated by a shadow deposition of metal atoms on a faceted NaCl (110) template. The microstructures of the nanowires are analyzed by means of transmission electron microscopy (TEM). The linear and nonlinear optical properties of the arrays of the nanowires are then investigated. Finally, the comparison between the experiments and the theories by Maxwell-Garnett model, and finite-difference time-domain (FDTD) method are discussed.

II. NANOWIRES FABRICATION

NaCl (110) single crystals having the dimension of 10 x 18 x 4 mm^3 are dipped in distilled water for 10 seconds and in ethanol for 1 minute in two cycles before being mounted on the substrate heating stage in an ultrahigh vacuum deposition chamber with a base pressure of approximately 2 x 10^{-9} Torr. The 200 - 300 nm thick homoepitaxial layers of NaCl are deposited after the substrate annealing at 200 - 450^{0}C. The 5 - 10 nm thick SiO layers are then deposited as passivation layers. Desired metal is deposited at the given deposition rate onto the substrate at the flux angle of 65^0 with respect to the

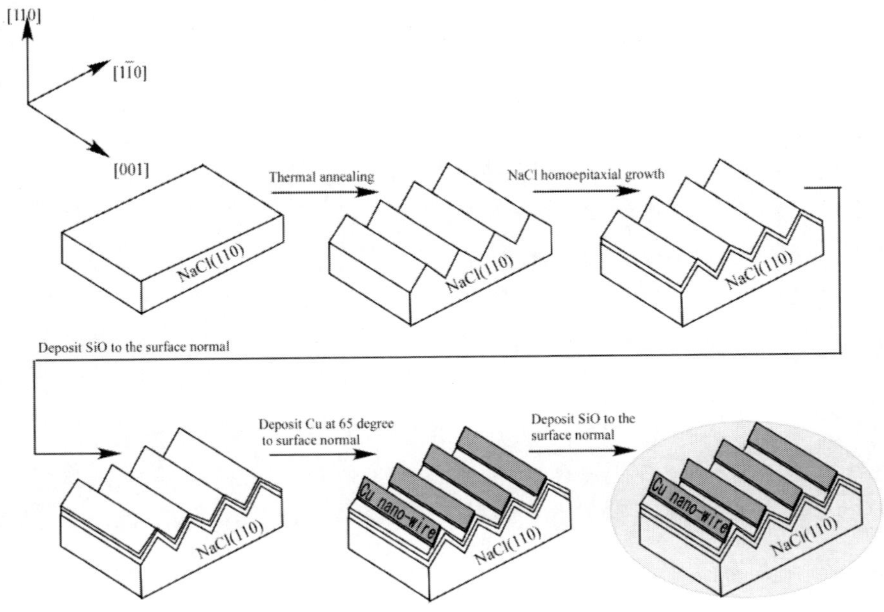

Figure 1. Schematic view of a shadow deposition method.

surface normal to form nanowires. On the arrays of metallic nanowires 5 - 10 nm thick SiO layers are deposited as protection layers. For transmission electron microscopy (TEM) observation, metallic nanowire arrays sandwiched by SiO layers are separated from the NaCl substrates by dissolving the substrates in distilled water, and mounted on 200 – 400 mesh Cu grids. Figure 1 gives a brief outline of the shadow deposition procedure.

III. Nanowires Characterization

Transmission electron microscopy (TEM) images in Figures 2 (a), (b), and (c) show the uniform Au, Cu, and Pt nanowires, respectively, as long and straight periodic dark images parallel to the [001] direction of the substrates. The nominal thicknesses of the nanowires are measured with the sensor plane of the thickness monitor set perpendicular to the metal beam direction and showed in Table 1. The minimum widths of the obtained nanowires are 40 nm for Au, 14 nm for Cu and 9 nm for Pt.

One of the key parameters determining the minimum width of the nanowires deposited onto a crystalline substrate is the surface energy of the wire metal (Figure 3). High surface energy of metals leads to high contact angle on the substrate surface as shown in Figure 4 [7]. As the surface energy

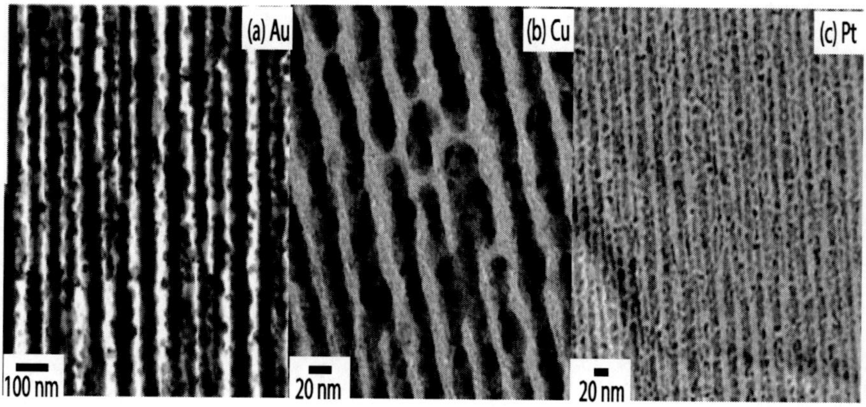

Figure 2. Transmission electron microscopy (TEM) images of Au (a), Cu (b) and Pt (c) nanowire arrays after the NaCl template has been dissolved. The metallic nanowire arrays are seen in dark contrast running parallel to the [001] direction of the NaCl (110) template. The minimum widths of the obtained nanowire arrays are 40 nm for Au, 14 nm for Cu and 9 nm for Pt.

Table 1. Geometrical parameters of the fabricated metallic nanostructures

Nanowires	Nominal thickness (nm)	Wire width (nm)	Periodicity (nm)
Au	65	40	80
Cu	10	14	27
Pt	7	9	17

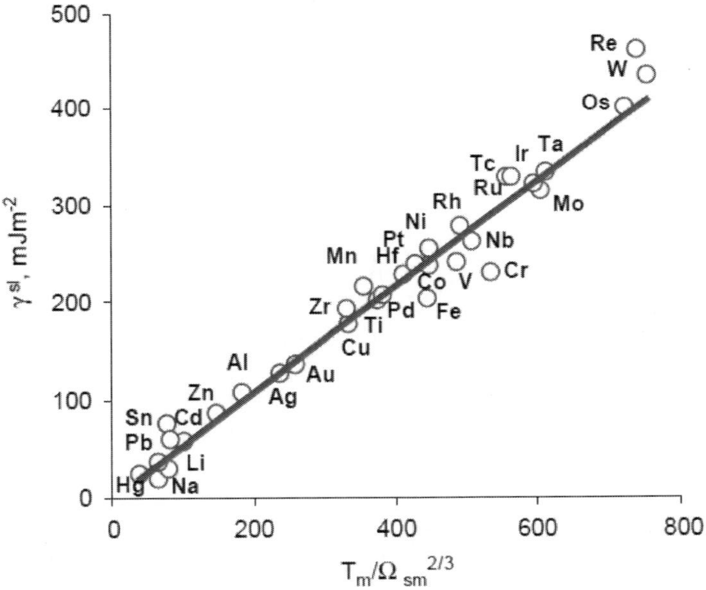

Figure 3. Surface energy of metals at melting point γ^{sl} with respect to $T_m/\Omega_{sm}^{0.7}$. T_m and Ω_{sm} are melting point and atomic volume of solid at melting point, respectively.

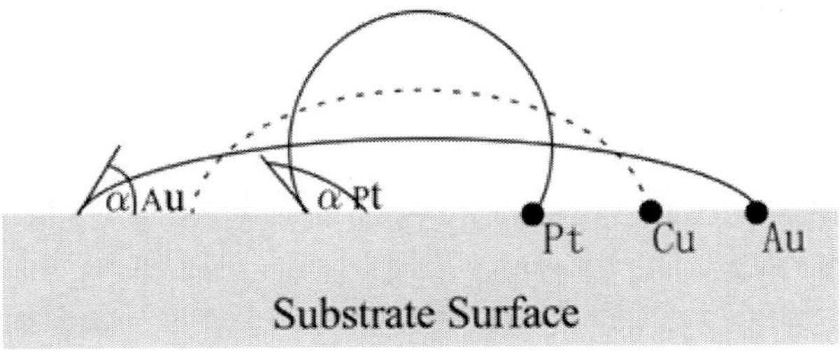

Figure 4. Equilibrium shape of a spherical drop from different contact angle of metals on flat solid substrate (α).

is the largest for Pt, the second largest for Cu, and the smallest for Au in their liquid state, the minimum width of Cu nanowires (14 nm) is larger than that of Pt (9 nm), but is smaller than that of Au (40 nm).

IV. Linear Optical Property

Reflectivity and absorption spectra can be measured as a function of the sample rotation angle. These spectra are recorded in the wavelength 250 - 1100 nm by a spectrometer equipped with a polarizer accessory, a Xenon lamp source, and a photomultiplier (Figure 5).

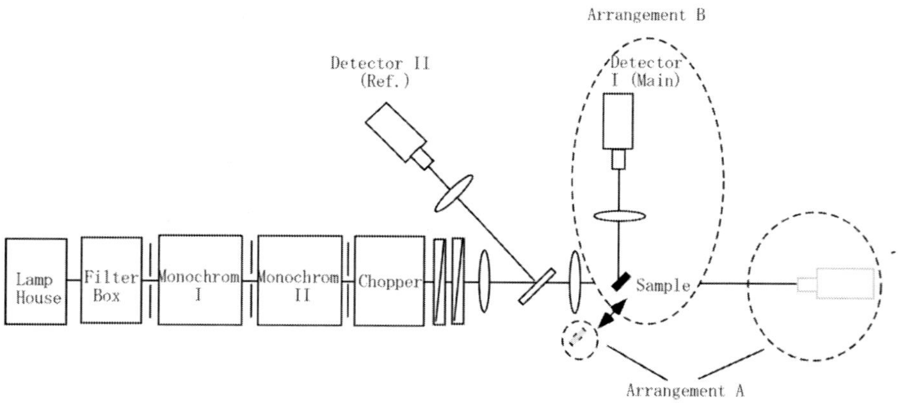

Figure 5. Schematic diagram of linear optical experimental setup. Arrangement A and B are applied to measure absorption and reflectivity spectra, respectively.

A drop in linear reflectivity spectra of Au nanowires is observed around $\hbar\omega = 1.7$ eV for the incident field perpendicular to the wire axes in Figure 6 (a), but it is not observed for incident field parallel to the wire axes. In Figure 6 (c) the drop in the reflectivity spectra near 4.2 eV for Pt nanowires is observed for the electric field polarized perpendicular to the wire axes and they shift to lower photon energy for the parallel polarization.

Cu nanowires exhibit reflectivity-polarization angle trends opposite to those of Pt nanowires. The drop in reflectivity near 2.2 eV is observed for the electric field polarized perpendicular to the wire axes in Figure 6 (b) and they shift to higher photon energy for the parallel polarization. Interestingly, the drop in reflectivity spectra near 2.2 eV for Cu nanowires in Figure 6 (b) appears to have a similar origin as the main absorption peak at 2.3 eV for Cu nanowires in Figure 7. Absorption maxima are remarkably observed near 2.3 eV for the perpendicular polarization in Figure 7 and they shift to lower energy for the parallel polarization. This possibly may be a plasmon excitation in the nanostructures giving rise to the local electric field enhancement.

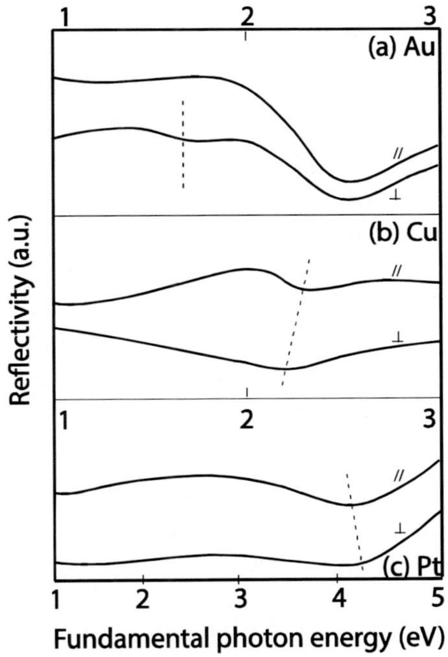

Figure 6. Measured linear reflectivity spectra of Au (a), Cu (b) and Pt (c) nanowire arrays. ⊥ (//) means the electric field perpendicular (parallel) to the wire axes.

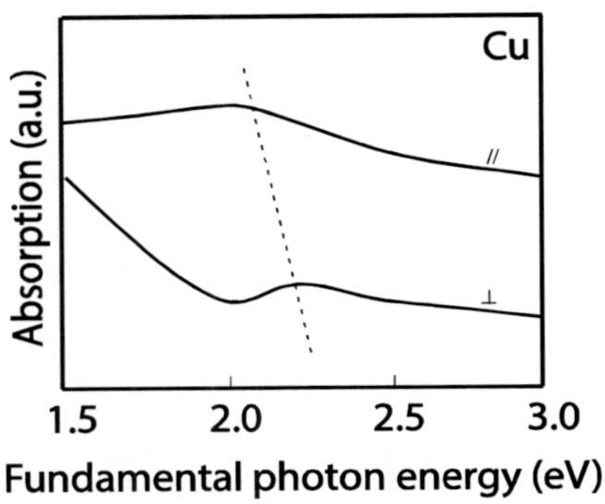

Figure 7. Measured linear absorption spectra of Cu nanowire arrays. ⊥ (//) means the electric field perpendicular (parallel) to the wire axes.

Theoretical analysis of linear reflectivity and absorption are performed in order to find whether the plasmon resonances served as the origin of the enhancement of the electromagnetic field from these metallic nanowires. The reflectivity and absorption of nanowires are calculated according to Maxwell-Garnett model [8-10]. For Cu nanowires, for instance, it is found that the detailed experimental dependence of the energy positions of the plasmon maxima on the polarization are different from the prediction by Maxwell-Garnett model [3]. When the electric field is perpendicular to the wire axes, the calculated positions of plasmon maxima are observed near 2.0 eV in contrast to the experiment showing plasmon maxima around 2.3 eV as seen in Figure 7. When the electric field is parallel to the wire axes, plasmon maximum energies are observed below 2.0 eV, and they conflict with the experiment in Figure 7 showing plasmon maxima above 2.0 eV. Another rigorous numerical study using the finite-difference time-domain (FDTD) method of solving the electromagnetic problems by integrating Maxwell's differential equations is then introduced [11, 12].

The FDTD method allows one to consider the actual wire size dependence of the peak energy positions for metallic nanowires [3]. This simulation is more favorable than that of Maxwell-Garnett model from a point of view of the reproduction of the experimental data. The calculation results show the plasmon maxima near 2.4 eV for Cu nanowires as the electric field is perpendicular to the wire axes, while the blue-shift occurs when the electric field is paralleled to the wire axes. The peak positions in the latter polarization configuration are above 2.0 eV, as is consistent with the experimental results in Figure 7. This present calculation taking into account the absolute wire size by FDTD simulation is judged to have improved the agreement between the experiment and the theory because it takes full account of the retardation effect.

From the discussion so far, the theoretical results indicate that the resonant enhancement near the fundamental photon energy of 1.7 eV for Au, 2.2 eV for Cu, and 4.2 eV for Pt possibly originate from the coupling of plasmon resonances in the metallic nanowires, however, these findings will be confirmed by their non-linear optical properties in the last section of this chapter.

V. Azimuthal Angle Dependence of Second Harmonic Intensity

The probe-light pulses with a fundamental photon energy of 1.17 eV for Au and Cu (2.23 eV for Pt) can be generated by a mode-locked Nd:YAG laser (Figure 8). The fundamental light beam is focused onto the sample surface and the incident angle is 45°. The second harmonic (SH) signal is detected by a photomultiplier and the absolute SH intensities are obtained by normalizing the SH response to that of the α-SiO_2 (0001).

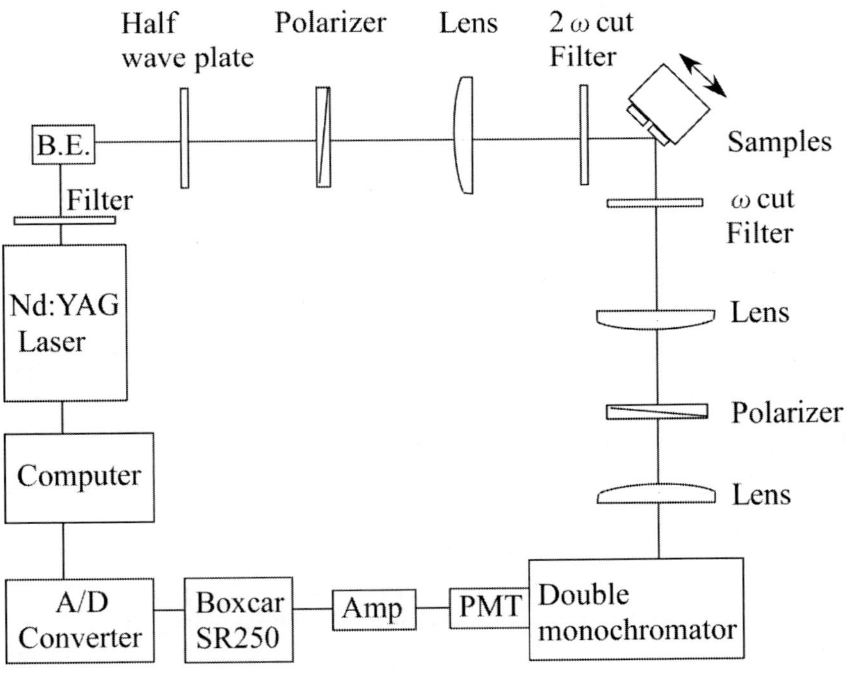

Figure 8. Block diagram of the experimental setup for measuring SHG rotational anisotropy. The fundamental laser beam is incident on the sample at an angle of 45°, and SH photons are collected in reflection as a function of the sample rotation angle ϕ. The ϕ is defined as the angle between the plane of incidence and the [001] direction of the sample surface. When $\phi = 270°$ the incident plane is defined in the same direction as the desired metal flux during the deposition. The rotational anisotropy in SH intensity patterns is obtained with different input and output polarization configurations. The B.E., PMT, Amp, and A/D stand for beam elevator, a photomultiplier, amplifier, and analogue-to-digital converter, respectively.

To measure the azimuthal angle dependence of the SH intensity, the sample is mounted on an automatic rotation stage with the surface normal set parallel to the rotation axis of the stage. The measurements are carried out in four different input/output polarization combinations: p-in/pout, p-in/s-out, s-in/p-out, and s-in/s-out. All SHG observations are carried out in air at room temperature.

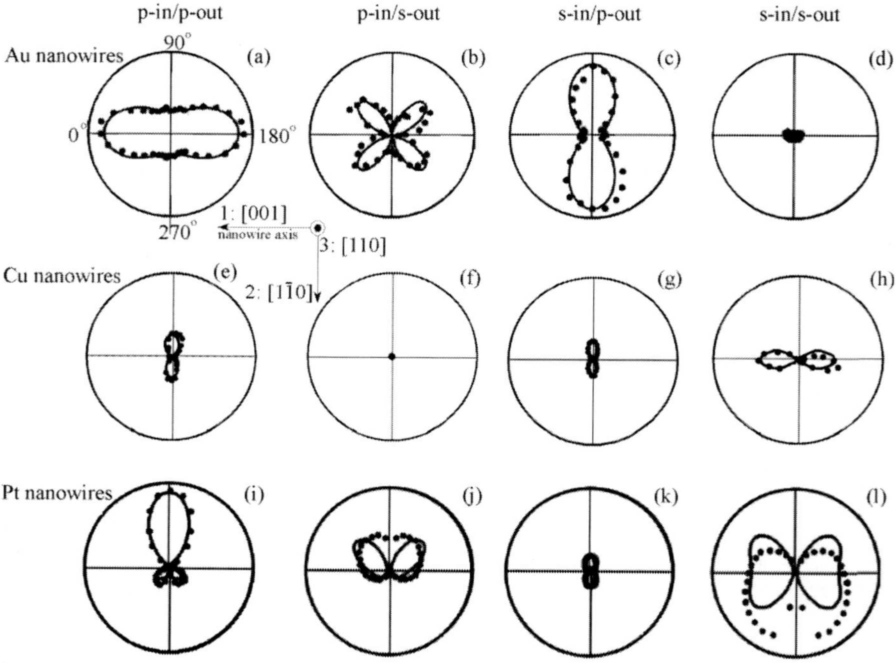

Figure 9. Measured (filled circles) and calculated (solid line) second harmonic (SH) intensity patterns for different metallic nanowire arrays as a function of the sample rotation angle ϕ. The fundamental photon energy is 1.17 eV for Au and Cu nanowire arrays, 2.23 eV for Pt nanowire arrays. The angle ϕ is defined as the angle between the incident plane and the [001] direction on the sample surface. The incident angle is 45°. The SH intensity is plotted in polar coordinates. The four different input and output polarization configurations are indicated at the top. The intensity scales are the same for the four polarization configurations for each sample.

Azimuthal angle dependence of the SH intensity from the Au, Cu, and Pt nanowire arrays is measured and showed in Figure 9. The SH intensity is measured as a function of the sample rotation angle ϕ at the photon energy of 1.17 eV for Au and Cu and at the photon energy of 2.33 eV for Pt. The fundamental output at 1.17 eV cannot be used for Pt nanowires because they are easily damaged due to their strong optical absorption. The angle ϕ is defined as the angle between the incident plane and the wire axes in the [001] direction. In each row the SH intensity patterns for four different configurations of input and output polarizations are showed for each sample.

The SH intensity from metallic nanowires depends strongly on the sample rotation angle ϕ and the polarization configurations of the fundamental and SH light. The patterns of the SH intensity from Au nanowires shows two lobes for the p-in/p-out configuration in Figure 9 (a), four lobes for the p-in/s-out polarization configuration in Fig. 9 (b), and two lobes for the s-in/p-out polarization configuration in Figure 9 (c). The SH intensity patterns from Cu nanowires display two main lobes for the p-in/p-out, s-in/p-out, and s-in/s-out polarization configurations in Figures 9 (e), (g), and (h), respectively. For Pt nanowires, two lobes are seen in any incident and output light polarizations as shown in Figures 9 (i)–(l).

The SH intensity patterns from these metallic nanowires in Figure 9 (filled circles) are analyzed by a phenomenological model [13, 14]. The fitted results are showed in the solid curves in Figure 9. The rotation-angle dependences of the mainly second-order susceptibility tensor elements for different metallic nanowires are then indicated in Table 2.

Table 2. Azimuthal-angle dependences of the mainly second-order susceptibility tensor elements for different metallic nanowire arrays

Nanowires	p-in/p-out	p-in/s-out	s-in/p-out	s-in/s-out
Au	$\chi_{113}^{(2)}, \chi_{223}^{(2)}, \chi_{333}^{(2)}$	$\chi_{113}^{(2)}, \chi_{223}^{(2)}$	$\chi_{311}^{(2)}, \chi_{333}^{(2)}$	
Cu	$\chi_{323}^{(2)}$		$\chi_{311}^{(2)}$	$\chi_{222}^{(2)}$
Pt	$\chi_{113}^{(2)}, \chi_{223}^{(2)}, \chi_{323}^{(2)}$	$\chi_{113}^{(2)}, \chi_{223}^{(2)}$	$\chi_{311}^{(2)}$	$\chi_{222}^{(2)}$

It is found that the main nonlinear susceptibility elements $\chi_{113}^{(2)}$, $\chi_{223}^{(2)}$, $\chi_{311}^{(2)}$, and $\chi_{333}^{(2)}$ dominate the SH signal for Au nanowires. The indices 1, 2, and 3 denote the [001], $[1\bar{1}0]$, and [110] directions on the NaCl (110) surface, respectively. The major contributions of the $\chi_{323}^{(2)}$, $\chi_{311}^{(2)}$, and $\chi_{222}^{(2)}$ elements are proved to be prominent the SH signal for Cu nanowires, whereas the main

$\chi^{(2)}_{113}$, $\chi^{(2)}_{223}$, $\chi^{(2)}_{222}$, $\chi^{(2)}_{311}$, and $\chi^{(2)}_{323}$ elements are dominant the SH signal for Pt nanowires. The positive and negative directions of the 2:$[1\bar{1}0]$ direction for Pt nanowires in Figure 9 (i) are not equivalent due to the geometrical nature of the shadow deposition.

The $\chi^{(2)}_{113}$ and $\chi^{(2)}_{223}$ elements dominate second order optical nonlinearity of Au nanowires for the p-in/s-out polarization configuration in Figure 9 (b) partly due to a strong depolarization field in the wires. This similar effect is observed for Pt nanowires in Figure 9 (j), but it is not for Cu nanowires in Figure 9 (f). The prominent $\chi^{(2)}_{113}$ and $\chi^{(2)}_{223}$ elements also lead to an enhancement of the optical nonlinearity of Au and Pt nanowires for the p-in/p-out polarization configuration. As clearly seen in this polarization configuration, the depolarization effect from Au nanowires in Figure 9 (a) is judged to be stronger than that from Pt nanowires in Figure 9 (i). The depolarization effect has rarely been observed as the origins of optical enhancement for Cu nanowires in Figure 9 (e). The two-lobed patterns in Au nanowires are wider in the middle in p-in/p-out (Figure 9 (a)) and s-in/p-out (Figure 9 (c)) polarization configurations due to the contribution from the $\chi^{(2)}_{333}$ element. Suppression of the $\chi^{(2)}_{113}$ and $\chi^{(2)}_{223}$ elements for Cu nanowires is existed due to a weak depolarization field in the wires leading to the intensity of SHG as low as the noise level for the p-in/s-out polarization configuration in Figure 9 (f).

The pattern for Cu nanowires for the s-in/s-out polarization configuration in Figure 9 (h) is seen to be dominated by the contribution of the $\chi^{(2)}_{222}$ element. This element results from the broken symmetry in the 2:$[1\bar{1}0]$ direction due to the cross-sectional shapes of the nanowires produced by the periodic macrosteps parallel to the 1:[001] direction. The similar effect is observed for Pt nanowires in Figure 9 (l). However it is not observed for Au nanowires in Figure 9 (d) because the incident electric field in the 2:$[1\bar{1}0]$ direction is partly canceled by the depolarization field perpendicular to the wire axes. This difference is partly due to the larger anisotropy in the nonlinearity of Cu and Pt nanowires due to thinner wires.

Local electromagnetic field is numerically investigated by FDTD method in order to clarify the contributions of the $\chi^{(2)}_{311}$ and $\chi^{(2)}_{323}$ elements dominated the optical nonlinearity of the nanowires [3]. It is found that the contribution of the $\chi_{323}^{(2)}$ element originating from local plasmons in Cu nanowires dominates the optical nonlinearity for the p-in/p-out polarization configuration in Figure 9 (e), and the contribution of the $\chi_{311}^{(2)}$ element originating from the lightning-rod effect dominates the optical nonlinearity for the s-in/p-out polarization configuration in Figure 9 (g). These similar effects also lead to an enhancement of the optical nonlinearity of Pt nanowires. The contribution of the $\chi_{323}^{(2)}$ element in Pt nanowires originating from the plasmon resonances dominates the optical nonlinearity for the p-in/p-out polarization configuration in Figure 9 (i), and the contribution of the $\chi_{311}^{(2)}$ element originating from the lightning-rod effect dominates the optical nonlinearity for the s-in/p-out polarization configuration in Figure 9 (k). For Au nanowires, it is possible that the contribution of the $\chi_{311}^{(2)}$ element originating from the lightning-rod effect dominated the optical nonlinearity for the s-in/p-out polarization configuration in Figure 9 (c). The contribution of the $\chi_{323}^{(2)}$ element originating from the plasmon resonances is not observed for the p-in/p-out polarization configuration in Figure 9 (a) due to a strong depolarization field in Au nanowires.

From the analysis of the SH intensity patterns by a phenomenological model and FDTD method it can be concluded that SHG from Cu (Pt) nanowires is found to be different from those of Au nanowires. This is due to the weaker suppression of the incident field by the depolarization field in Cu (Pt) nanowires and the stronger anisotropy in the nonlinearity of Cu (Pt) nanowires due to thinner wires than Au. The electric field component along the surface normal enhanced by the plasmon-resonant coupling in the wires is observed for Cu (Pt) nanowires, but it is not easily noticed for Au nanowires.

VI. Wavelength Dependence of Second Harmonic Intensity

Light source of the fundamental photon energy from 1.2 to 2.3 eV can be an optical parametric generator and amplifier system driven by a mode-locked Nd:YAG laser (Figure 10). The excitation light is focused on the sample surface at the incident angle of 45°. The SH signal from the sample is detected by a photomultiplier. Finally, the absolute magnitudes of the SH intensity are obtained by normalizing the SH response to that of the α-SiO_2 (0001). The measurements are performed in p-in/p-out polarization configuration. All SHG observations are performed in air at room temperature.

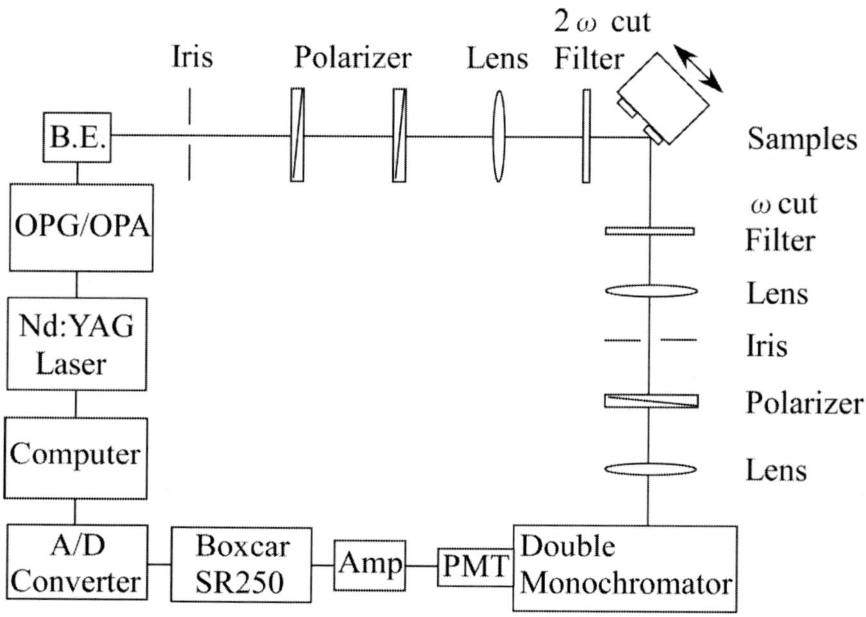

Figure 10. Block diagram for measuring the SH intensity as a function of the photon energy. The wavelength tunable light source is the optical parametric generator and amplifier (OPG/OPA) system and it is incident on the sample at an angle of 45 °. The B.E., PMT, Amp, and A/D represent beam elevator, a photomultiplier, amplifier, and analogue-to-digital converter, respectively.

The SH intensity spectra from metallic nanowires are measured as a function of the SH photon energy for the p-in/p-out polarization combination as shown in Figure 11. The incident wave vector of the excitation is parallel to the $[1\bar{1}0]$ direction of the NaCl (110) faceted substrates. The SH intensity spectra taken from Au nanowires in Figure 11 (a) show a nearly flat dependence of SH intensity on the photon energy below $2\hbar\omega = 3.0$ eV. The intensity abruptly increases above $2\hbar\omega \sim 3.1$ eV, and reached a maximum at 3.3 eV. The SH intensity spectra for Cu nanowires in Figure 11 (b) showed steady increase above $2\hbar\omega = 3.0$ eV. The SH response exhibits a peaked resonance near 4.4 eV. In Figure 11 (c) the SH intensity spectra for Pt nanowires show steady increase above $2\hbar\omega = 3.7$ eV. The SH response exhibits a peaked resonance near 4.3 eV.

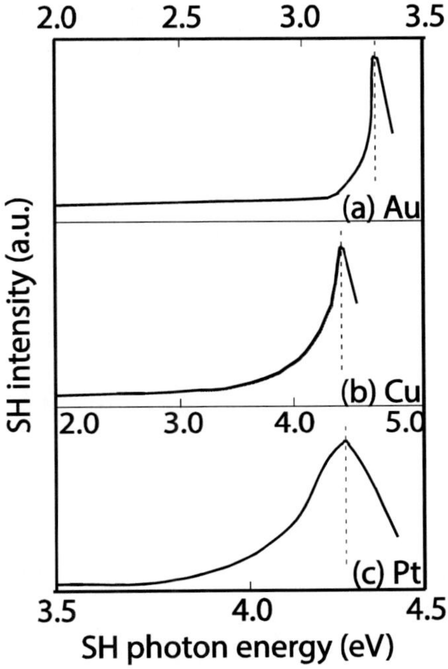

Figure 11. Measured second harmonic (SH) intensity spectra from different metallic nanowire arrays as a function of the SH photon energy for the p-polarized input and p-polarized output configuration. The incident wave vector of excitation is parallel to the [1$\bar{1}$0] direction of the NaCl (110) faceted substrates.

The main peak near $2\hbar\omega = 3.3$ eV for Au nanowires in Figure 11 (a) appears to have a similar origin as the main peak at 4.4 eV for Cu nanowires in Figure 11 (b), and the main peak at 4.3 eV for Pt nanowires in Figure 11 (c). Two candidate origins of the SHG from these metallic nanowires are considered. They are (1) the enhancement of the local field by the plasmon excitation in the nanowires, and (2) the dielectric response at the metal/SiO interface.

Let us discuss a resonance enhancement of the SH signal with the dipolar mode of the surface plasmon in the metal wires. Theoretical analysis of the SH intensity is performed in order to determine the values of the local field factor and then to find whether the surface plasmon resonance served as the origin of the enhancement of the SH response from the metallic nanowires. Local field enhancement factor for Au, Cu, and Pt nanowires is calculated by a quasi-static theory [9]. The calculated enhancement factor is described in reference [3, 15, 16]. It is found that the calculated spectra for the metallic nanowires in Figure 12 reproduce the experimental findings in Figure 11 well. Hence, these theoretical results indicate that the resonant enhancement near the SH photon energy of 3.3 eV for Au, 4.4 eV for Cu, and 4.3 eV for Pt originate from the coupling of plasmon resonances in the metal nanowires. Experimental findings and calculations are consistent with the suggestion by Schider et al. that the plasmon excitation in metallic nanostructures giving rise to the local electric field enhancement [2]. For Au nanowires, it is noted that the nonlinear response was weak at the SH photon energy lower than 3.3 eV, due to the anisotropic depolarization field.

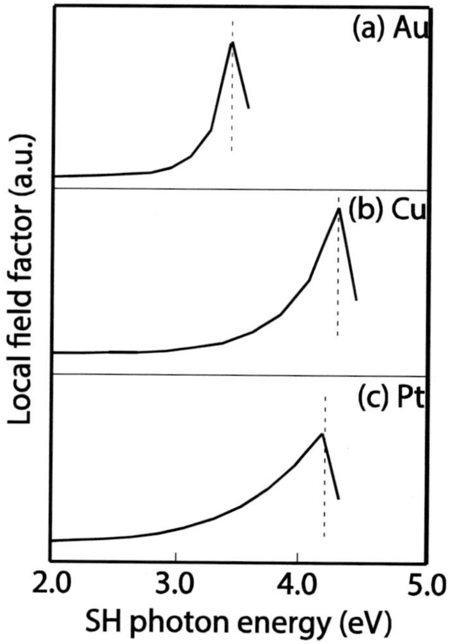

Figure 12. Calculated second harmonic generation (SHG) local field enhancement factor as a function of the SH photon energy for different metallic nanowire arrays according to reference [9].

Next, we consider an enhancement of the optical nonlinearity of metallic nanowires assigned to the electronic resonance at the metal/SiO interface. For Au nanowires, the SH intensity spectrum has a main peak around $2\hbar\omega = 3.3$ eV in Figure 11 (a). It is suggested that the Fresnel factor for the incident light creates a structure at 1.7 eV due to the linear dielectric structure seen in the linear reflectivity in Figure 6 (a), and enhances the SH intensity at twice this photon energy. Like Au nanowires, the SH intensity spectrum of Cu nanowires has a main peak around $2\hbar\omega = 4.4$ eV in Figure 11 (b). It is also suggested that the Fresnel factor for the incident light creates a structure at 2.2 eV due to the linear dielectric structure seen in the linear reflectivity in Figure 6 (b), and enhanced the SH intensity at twice this photon energy. In contrast, the drop in linear reflectivity spectra for Pt nanowires is observed near 4.2 eV in Figure 6 (c). It is possible that a dielectric structure exists near $2\hbar\omega = 4.3$ eV for Pt nanowires in Figure 11 (c) and the Fresnel factor at frequency 2ω for SHG is enhanced by one photon resonance of SH photons. Hence, these findings clearly suggest that the observed structures in the SH spectra of metallic nanowires is explained as a result of the linear dielectric response of the composites of the metallic nanowires in SiO matrix. They lead to an enhancement of second order optical nonlinearity of metallic nanowires.

From the analysis of the SH intensity by a quasi-static theory one can conclude that the SHG from metallic nanowires is originated from the plasmon resonance and the linear dielectric properties of the metal/SiO interface.

Conclusion

Microstructures and optical properties of self-organized arrays of the metallic polycrystalline nanowires: Au, Cu and Pt are studied. Shadow deposition onto (110) faceted faces of NaCl crystals is demonstrated as a simple technique for fabricating wide-area arrays of the nanowires suitable for both transmission electron microscopy and optical measurements. It is found that there are two candidate origins of the observed SH intensity from metallic nanowires. One is the enhancement of the local field by the plasmon excitation in the nanowires. Another one is the dielectric structure at the metal/SiO interface. The second harmonic generation (SHG) from Cu (Pt) nanowires is found to be different from those of Au nanowires. This is due to the weaker suppression of the incident field by the depolarization field in Cu (Pt) nanowires and the stronger anisotropy in the nonlinearity of Cu (Pt) nanowires due to thinner wires. This difference may have significant consequences in the range of applications that such materials may be useful for.

References

[1] Sugawara A., Hembree G.G., Scheinfein M.R. *Journal of Applied Physics*. 1997, 82, 5662-5669.

[2] Schider G., Krenn J.R., Gotschy W., Lamprecht B., Ditlbacher H., Leitner A., Aussenegg F.R. *Journal of Applied Physics.* 2001, 90, 3825-3830.

[3] Locharoenrat K., Sugawara A., Takase S., Sano H., Mizutani G. *Surface Science.* 2007, 601, 4449-4453.

[4] Shen Y.R. The Principles of Nonlinear Optics; *Wiley:* New York, 2002, pp 5-50.

[5] Bloembergen N. Nonlinear Optics; World Scientific: *Miami,* 2005, pp 25-77.

[6] Boyd R.W. Nonlinear Optics; Academic: *Orlando,* 2008, pp 55-102.

[7] Digilov R.M. *Physica B.* 2004, 352, 53-60.

[8] Maxwell-Garnett J.C. *Philosophical Transaction*s of the Royal Society London A. 1904, 203, 385-420.

[9] Al-Rawashdeh N.A.F., Sandrock M.L., Seugling C.J., Foss J.C.A. *Journal of Physical Chemistry B.* 1998, 102, 361-371.

[10] Hayashi N., Aratake K., Okushio R., Iwai T., Sugawara A., Sano H., Mizutani G. *Applied Surface Science.* 2007, 253, 8933-8938.

[11] Sullivan D.M. Electromagnetic Simulation Using the FDTD Method; *IEEE:* New York, 2013, pp. 7-105.

[12] Taflove A. Computational Electrodynamics: The Finite-Difference Time-Domain Method; *John Wiley & Sons:* New Jersey, 2010, pp 13-108.

[13] Sipe J.E., Moss D.J., van Driel H.M. *Physical Review B.* 1987, 35, 1129-1141.

[14] Kobayashi E., Mizutani G., Ushioda S. *Japanese Journal of Applied Physics.* 1997, 36, 7250-7256.

[15] Kitahara T., Sugawara A., Sano H., Mizutani G. *Journal of Applied Physics.* 2004, 95, 5002-5005.

[16] Takabe H., Luong N.H., Onuki Y. Frontiers of Basic Science Towards New Physics, Earth and Space Science, Mathematics; Osaka University Press: Osaka, 2006, pp 53-104.

In: Nonlinear Optics
Editor: Fred Wilkins

ISBN: 978-1-63321-926-7

Chapter 6

NONLINEAR OPTICAL PROPERTIES IN IONIC LIQUIDS

M. Trejo-Durán*[*]*, E. Alvarado-Méndez and J. M. Estudillo-Ayala

División de Ingenierías Campus Irapuato Salamanca
Universidad de Guanajuato, México

ABSTRACT

Nowadays it is interesting the huge amount of ionic liquids research that involves their green chemistry and their physical and chemical properties, whilst there has not been a deep research involving their potential applications in optical properties and optoelectronic devices. In this chapter, we show a review of recent research results on this topic, from nonlinear optical properties to potential applications; theoretical and experimental results of common ionic liquids. We also present new experimental results in NLO properties of six ionic liquids. Finally, we summarize some ionic liquids applications in the optical and optoelectronic areas.

[*] Corresponding author: email: mtrejo@ugto.mx.

I. Introduction

Currently there is a great number of materials which present nonlinear optical properties, some of them are reported in [1-4] but there are few reports on ionic liquids [IL] with nonlinear optical properties. Room Temperature Ionic Liquids (RTILs) are salts composed by bulky organic cations and non-coordinating inorganic anions; they present an ionic-covalent molecular structure and different molecular architectures. They have shown to be excellent solvents for optical investigations of solutes at room temperature with low environmental impact. The IL possesses low melting point temperatures (below 373°K), good electrical conductivity, high ionic mobility, negligible vapor pressure, and excellent chemical and thermal stability. Most of the commonly used ionic liquids are semi-transparent in the UV (>300 nm) and completely transparent in the visible region of light. The most common ionic liquids are based on imidazolium cation, the reason for this is that they are the most studied in many areas. An important feature of IL are their optical properties, these can be linear and nonlinear. For example fluorescence is a linear property and can be applied to sensors design; the fluorescence behavior of some IL depends strongly on the excitation wavelength. However, their nonlinear optical applications are currently unexplored.

The aim of this chapter is to give an overview about nonlinear optical properties of ionic liquids from experimental methods to numerical calculus including molecular and microscopic point of view. Finally, we talk about some potential applications for photonic devices.

At the end of the chapter, (Table 4 and Table 5), we enclose a list of nomenclature and the names of cations and anions of ionic liquids mentioned here.

II. Brief Nonlinear Optical Properties Theory

Light-matter interaction in optics, describes the relation between an electro-magnetic wave and a material, in which the molecular system resides. In the presence of external electric field, all charged particles in the molecule, electrons as well as the atomic nuclei, will experience a force that disrupts their motions. The oscillating charges in the molecule give rise to an induced dipole moment (μ), and the degree to which the external field $\boldsymbol{E(t)}$ manages to set the charges in motion is expressed as Taylor series

$$\mu(t) = \mu^0 + \alpha E(t) + \tfrac{1}{2}\beta E^2(t) + \tfrac{1}{6}\gamma E^3(t) + \cdots \quad (1)$$

Where μ^0 is the permanent electric dipole moment, α the linear electric polarizability, β the fist-order hyperpolarizability, γ the second-order hyperpolarizability, and so forth. In optics, this equation provides the fundamental origin of nonlinearities, and, at the microscopic level, the hyperpolarizabilities govern the nonlinear optics properties performance of the material.

Induced dipolar moment (μ) is a main molecular parameter to analyze due to the fact that it represents all responses, linear and nonlinear, to an external applied electric field (***E***, DC or electromagnetic radiation). The linear response of a material is present by changes in linear electric polarizability (α) while nonlinear response is present by changes at hyperpolarizability of first (β) and second (γ) order.

In macroscopic case, dielectric materials are characterized by their polarization ***P*** *(t)* of the medium and, if it expands the polarization in orders of the applied external field we have

$$P(t) = P_L + P_{NL} \quad (2)$$

$$P_{NL} = \chi^{(2)}.EE + \chi^{(3)}.EEE + \cdots \quad (3)$$

Where $\chi^{(2)}$ and $\chi^{(3)}$are the quadratic and cubic susceptibility parameters respectively that determine the magnitude of the second and third order nonlinear response whose typical phenomena are second -harmonic generation and Kerr (optic and electrooptic) effect respectively [5,6].

In general, third-order optical nonlinear susceptibility $\chi^{(3)}$ is a complex quantity.

In reference to Kerr optic effect, this is present in a material in which third-order optical nonlinear susceptibility $\chi^{(3)}$ is non zero, and we can observe their presence in the refraction index changes. These changes can be originated by electronic polarization, molecular orientation, thermal effects, electrostriction, or saturated atomic absorption [7].

Changes in refraction index can be expressed by:

$$n = n_0 + \Delta n \quad (4)$$

And Δn origin, depending on the experimental conditions, can be due to one or multiple mechanisms; one of them is the electronic response of the medium when the rise-up time is shorter than the duration of pulses delivered by the laser source used in the experiments, the other is the thermal effect which is caused by the heating of the medium due to a linear or nonlinear absorption process.

For electronic response (electronic Kerr nonlinearity), the change in the refraction index depends on the optical irradiance,

$$n = n_0 + n_2 I \tag{5}$$

Where n_2 is the nonlinear refraction index, and it is proportional to $\chi^{(3)}$ and is related by [8]:

$$n_2 = \frac{3\eta_0}{n^2 \epsilon_0} \chi^{(3)} \tag{6}$$

where $\eta_0 = (\mu_0/\varepsilon_0)^{1/2}$, n is the refraction index, ε_0 is the permittivity in free space.

In reference to thermal response of the materials which have low absorption, the change in refractive index is given by:

$$n = n_0 + \frac{dn}{dT} \Delta T \tag{7}$$

Where $\frac{dn}{dT}$ is the thermo-optic coefficient and ΔT is the nonlocal temperature change which depends on the optical fluence [9]

The main difference between this two response is that the electronic response, is local and fast, originated by distortions in electronic cloud due to the electric field which induces changes in optical susceptibility of the medium while the thermal response, is nonlocal, slow and accumulative, originated by heat conduction.

Nonlinear absorption is other interesting property, the absorption depends on the irradiance and nonlinear absorption coefficient (β) which are related by the equation:

$$\alpha(I) = \alpha + \beta I \tag{8}$$

where α is the linear absorption coefficient and I the irradiance of the laser beam.

Nonlinear absorption (NLA) mechanisms are present in one of two forms: Saturable Absorption (SA) or Reverse Saturable Absorption (RSA), RSA that could have been originated by Two-Photon Absorption (TPA), Excited State Absorption (ESA), Free Carrier Absorption (FCA), nonlinear scattering or with the combination of these processes [10-12].

III. NONLINEAR MEASUREMENT TECHNIQUES

Some common methods to determine nonlinear optical response at microscopic and macroscopic view are respectively, Optically Heterodyned Optical Kerr effect (OHD-OKE), and Z-scan among others.

III.1. Optically Heterodyned Optical Kerr Effect (OHD-OKE)

In general, third-order optical nonlinear susceptibility $\chi^{(3)}$ is a complex quantity. However, for the majority of photonics applications, such as all-optical switching, signal processing, and data storage, the real component of $\chi^{(3)}$ is more relevant. Unfortunately, in conventional optical Kerr effect (OKE) experiments, the detected signal is proportional to the quadratics of the third-order optical nonlinear susceptibility $\chi^{(3)}$. Hence, the real components of $\chi^{(3)}$ cannot be determined directly from the experiment but, by the introduction of local oscillator field with or without 90° optical phase bias with respect to the optical Kerr signal, a new experimental technique was developed, the optically heterodyned optical Kerr effect (OHD-OKE) [13,14], this new technique allows the determination of the magnitudes and signs of both the real and the imaginary components of $\chi^{(3)}$ with greatly improved signal-to-noise ratios.

III.2. Z-Scan Technique

Z-scan is simple but useful technique that allows to obtain the nonlinear refraction index and nonlinear absorption in only one measurement each.

Bahae developed this technique and their formulism [15], which use a laser beam, a lens aligned with the laser, a sample located in way of the laser after the lens, and a photodetector (example see Figure 1). The mechanism to obtain the information is simple; the sample is displaced along "*z* axis" across focus of lens while the transmitted light is collected in photodetector. It gives information on transmittance vs displacement "*z*" and, the curve obtained is characteristic of this technique. Nonlinear absorption characteristic curve using an open aperture and nonlinear refractive characteristic curve using a close aperture in the photodetector can be obtained respectively. Information of NLO properties of the sample can be obtained using different models [15-24] only two are briefly mentioned here:

III.2.1. Bahae model

The normalized transmittance for close aperture z-scan, is given as:

$$T(z) = 1 + \Delta\Phi_0 \frac{4x}{(1+x^2)(9+x^2)} \tag{9}$$

here $\Delta\Phi_0 = \frac{2\pi n_2 I_0 L_{eff}}{\lambda}$ is the maximum change phase shift on axis at focus, x is the dimensionless sample position, $x = \frac{z}{z_0}$, $z_0 = \frac{\pi\omega^2}{\lambda}$ is the Rayleigh parameter of the Gaussian beam with waist ω_0, n_2 is the nonlinear refraction index, $L_{eff} = (1-e^{-\alpha L})/\alpha$, is effective length and I_0 is the irradiance with the sample at the focus ($z=0$).

The normalized transmittance for open aperture z-scan, is given as:

$$T(z) = \frac{\ln[1+q_0(z)]}{q_0(z)} \tag{10}$$

For $|q_0(z)| < 1$, where $q_0(z) = I_0\beta L_{eff}/(1+\frac{z^2}{z_0^2})$. β is the nonlinear absorption coefficient.

III.2.2. Thermal Lens Model

Aparabolic approximation to the temperature field and an approximation of the optical effects due to temperature, give a behavior to the simple lens [17,18]. The time dependent expression [6] for the far- field on-axis transmittance is:

$$T_N = \frac{I(z,t)}{I(z,0)} = \frac{1}{1+\left(\frac{\theta}{1+t_c/2t}\right)\left(\frac{2x}{1+x^2}\right)+\left(\frac{\theta}{1+t_c/2t}\right)^2\left(\frac{1}{1+x^2}\right)} \quad (11)$$

Where $x = \frac{z}{z_0}, z_0 = \frac{\pi\omega^2}{\lambda}, t_c = \frac{\omega^2}{4k/C_p}$ t_c is the thermal diffusion time, k is the thermal conductivity and C_p is the specific heat per volume at constant pressure. The on-axis phase shift is given by $= \frac{\alpha PL}{\lambda\kappa}\frac{\partial n}{\partial T}$, where P is the power of the Gaussian beam and T is the temperature. To lower power the nonlinear phase shifts $\theta \ll 1$, then the second term of Eq. (5) considering $t >> t_c$ is given by

$$T_N = \frac{1}{1+(\theta)\left(\frac{2x}{1+x^2}\right)} \quad (12)$$

IV. Nonlinear Optical Properties. Experimental Data

IV.1 Nonlinear Refraction Index

This section present the experimental data about nonlinear properties of IL as n_2, $\frac{dn}{dT}$, and $\chi^{(3)}$. Studies about nonlinear refraction index of ILs, principally based on imidazolium cation, have been reported. Z-scan technique is the more popular technique to obtain the thermal lens strength (θ), thermo-optical coefficient ($\frac{dn}{dT}$) and nonlinear refraction index n_2.

Experimental results for ionic liquids where the n_2, or dn/dT were determined, have been reported [25-33], a summary of determined values and experimental conditions are show in Table 1.

Souza et al. [25], studied two ionic liquids 1-n-butyl-3-methylimidazolium tetrafluoroborate ([C_4mim][BF4]) and 1-n-butyl-3-methylimidazolium hexafluorophosphate ([C_4mim][PF6]) using two different lasers, they conclude that origin of nonlinearities is thermal and does not present nonlinear absorption. They reported that different ionic liquid structures, determined by the anion, influence the nonlinear refractive index and thermo-optical coefficient dispersion.

Del Sesto et al. [26-28] studied nonlinear optical properties of ionic liquids with tetraalkylphosphonium, $[PR4]^+$ cation, with $[DTMN]^{2-}$, $[Co(NCS)_4]^{2-}$, $[xan]^-$, $[CoB]^-$ and $[DDTC]^-$ anions, also theoretical calculus of γ anions and cations. They explain that $[PR4]_2[DTMN]$ has a better nonlinear optical response than [PR4][DDTC] because it has more sulfur atoms and it is a dianion, as opposed to the DDTC monoanion. The increased response could also be due to having the stronger -CN electron withdrawing group at the opposite end of the anion.

Other researches about NLO in ILs have focused in the analysis of how the cation or anion in ionic liquids develop changes in their linear and nonlinear optical properties that will be discussed in following sections.

Effects of Anion and Cation in Nonlinear Optical Properties $(n_2, \frac{dn}{dT})$

As it is well known, increasing the length of N-alkyl substituents at the imidazolium ring of the cation results in an increase in the viscosity of the liquid at any particular temperature, as well as generally increasing the hydrophobic nature of the salt. The introduction of a methyl group at the 2-position on the ring results in a considerable decrease in the polarity of the ionic liquid. Different anions can also give room-temperature ionic liquids of distinctly different characteristics regarding to properties such as viscosity and miscibility with other solvents. Presence of different anion or cation on ionic liquid changes the nonlinear optical properties.

Santos et al. [29,32] studied nonlinear optical properties of ionic liquids with cation using five different derivatives of imidazolium ($[C_n mim]$, n=4,6,7,10) in two regions: infrared and violet. In infrared region they reported that the ILS with $[BF_4]^-$ presented the largest diffusivity while the $[Tf_2N]^-$ gave the smallest result. Similarly, the thermal nonlinear response does not change significantly with cation changes. Nevertheless, the compounds with $[Tf_2N]^-$ have a smaller thermal nonlinear refractive index compared with ILs composed by $[BF_4]^-$ and $[PF_6]^-$ anions in the near infrared region. The thermo-optical coefficient, and all the measured values, are contained in the range of $10^{-4} - 10^{-5}$ K^{-1}. They reported that $[C_4mim][BF_4]$ presents the largest value in modulus, while the $[C_6mim]$ and $[C_{10}mim]$ have the smallest results. It was also observed that structures with $[Tf_2N]^-$ as the anion, present lower nonlinear response in comparison with other ILs with the same cation. Moreover, for a fixed anion $[Tf_2N]^-$ changing the cationic part does not significantly modify the value of the thermal contribution to n_2. In the purple region they compared the same ILs with the anion $[Tf_2N]^-$ using the chain of imidazolium $[C_nmim]^+$ (n=4,6,8,10), they observed that the cationic changes modify the value of the thermal contribution to n_2 from -1.8 to -5.1 x10^{-8} cm^2/W and dn/dT from -2.5 to -12.2 x10^{-5} K^{-1}.

Novoa-Lopez et al.[30] reported their results of ILs (methyl-imidazolium, pyridinium families) analysis about thermal refraction; they mentioned, it is highly sensitive to anionic part of the IL, at reference to thermal conductivity and, confirm that the nature of the cation has only a slight influence on the thermal conductivity of the IL, Furthermore, the anion exerts the strongest influence on physical properties, as it has been observed. In particular $[BF_4]^-$ gives rise to higher thermal conductivities(thermo-optic coefficient) than $[Tf_2N]^-$ when combined with the same cation, furthermore, while longer is chain (n) in cation $[C_nmim]^-$ with the $[BF_4]^-$ anion, the thermo-optic coefficient increases, not as $[Tf_2N]^-$ anion.

Tran et al. [31] studied the thermal lens sensitivity, that is directly related to thermal lens strength (θ) [34] and thermo-optical coefficient ($\frac{dn}{dT}$), to ILs with $[Tf_2N]^-$ anion and different cations as $[C_nmim]^+$ with n=2,4,5,6,8 and different anion as $[BF_4]^-$, $[PF_6]^-$ and $[Tf_2N]^-$. They found that θ is higher in ion $[PF_6]^-$ and lower in $[Tf_2N]^-$ using the same $[C_4mim]^+$ cation, respect to the same anion the results shown they found that increasing the chain length of the alkyl group on the imidazolium cation led to an increase in the sensitivity, which means θ increase.

In summary, the works that have studied the anion and cation effects over the nonlinear optical properties in ionic liquids, match in that the ionic liquids with cation imidazolium, present a higher nonlinearity when the anion is $[BF_4]$ than when the anion is $[Tf_2N]^-$, whilst in some cases the changes in the (n) cation chain length with the same anions, increase the nonlinearity.

Valencia et al. [33], used four ionic liquids with same cation methylpyrrolidine and different anion as $[HNO_3]$, $[HSO_4]$, [HCOO] and $[CH_3COO]$, they found that $[CH_3COO]$ presents a good nonlinearity compared to other anions. Their values are bigger than others reported to dyes by [35].

Other studies about cation's effect were reported by Xia [36] who reports the effect of symmetry on the low-frequency spectra of imidazolium ionic liquids.

Table 1. Experimental values of n_2 or *dn/dT to several ionic liquids*

Ionic liquid	α_o (cm^{-1})	Characteristics laser $\lambda-\tau-\nu$	$-dn/dT$, (10^{-5} °K^{-1})	$-n_2$ (10^{-10} cm^2/W)
[C_2mim] [BF_4] [30]	0.081	810nm-80 fs-80.75 MHZ	1.1	NA
[C_2mim] [BF_4][37]	0.25	514nm-CW-Argon	NA	45.5
[C_2mim] [Tf_2N][30]	0.173	810nm-80 fs-80.75 MHZ	0.4	NA
[C_2mim] [Tf_2N][37]	0.019	514nm-CW- Argon	NA	25
[C_2mim] [$EtSO_4$] [30]	0.025	810nm-80 fs-80.75 MHZ	2.8	NA
[C_4mim] [BF_4]	0.015	810nm-80 fs-80.75 MHZ	4.8[30], 75.4[31]	NA
[C_4mim] [Tf_2N]	0.049	810nm-80 fs-80.75 MHZ	1.1[30], 40.2[31]	NA
[C_4mim] [Tf_2N][37]	0.0194	514nm-CW- Argon	NA	33.9
[C_4mim] [BF_4][25]	0.014	810nm-200 fs-76 MHZ	20±6	18±5
[C_4mim] [BF_4][25]	0.29,	514nm-CW- Argon	104	28
[C_4mim] [PF_6][25]	0.032	810nm-200 fs-76 MHZ	8.1±2.4	27±8
[C_4mim] [PF_6][25]	0.054	514nm-CW- Argon	105	97
[C_4mim][Tf_2N] [29]	0.005	820nm-200 fs-76 MHZ	6.0±1.8	4.0±1.2
[C_4mim][Tf_2N] [32]	0.161	410nm-200 fs-(100Hz-1KHz)	9.6±2.9	390±120
[C_6mim] [Tf_2N][30]	0.059	810nm-80 fs-80.75 MHZ	0.9	NA
[C_6mim] [Tf_2N] [29]	0.012	820nm-200 fs-76 MHZ	3.0±0.9	4.9±1.5
[C_6mim] [Tf_2N][32]	0.071	410nm-200 fs-(100Hz-1KHz)	9.3±2.8	190±60
[C_6mim] [BF_4][30]	0.03	810nm-80 fs-80.75 MHZ	25	NA
[C_8mim] [Tf_2N][29]	0.007	820nm-200 fs-76 MHZ	7.0±2.1	5.8±1.7
[C_8mim] [Tf_2N][32]	0.240	410nm-200 fs-(100Hz-1KHz)	12.2±3.7	510±150
[C_8mim] [BF_4][30]	<0.03	810nm-80 fs-80.75 MHZ	34	NA
[C_{10}mim.] [Tf_2N][29]	0.012	820nm-200 fs-76 MHZ	3.0±0.9	4.7±1.4
[C_{10}mim] [Tf_2N][32]	0.319	410nm-200 fs-(100Hz-1KHz)	2.5±.08	180±50

Ionic liquid	α_o (cm^{-1})	Characteristics laser $\lambda-\tau-\nu$	*-dn/dT*, (10^{-5} °K^{-1})	$-n_2$ (10^{-10} cm^2/W)
[C_{10}mim] [BF_4][30]	0.03	810nm-80 fs-80.75 MHZ	45	NA
[C_{12}mim] [Tf_2N][29]	0.013	820nm-200 fs-76 MHZ	NA	5.0±1.5
[C_4mim] [MetFSO4] [30]	0.043	810nm-80 fs-80.75 MHZ	1.9	NA
[C_4C_1Pyrr] [Tf_2N][30]	0.039	810nm-80 fs-80.75 MHZ	1.3	NA
[C_6C_6 C_6C_{14}P] [Tf_2N][30]	0.074	810nm-80 fs-80.75 MHZ	1.5	NA
[C_2mim] [CF_3COO][37]	0.08	514nm-CW- Argon	NA	31.3
[C_4mim] [CF_3COO][37]	0.08	514nm-CW- Argon	NA	18
[C_1C_1Pyrr] [NO_3] [33]	0.95	514nm-CW- Argon	NA	1469
[C_1C_1Pyrr] [HSO_4] [33]	0.20	514nm-CW- Argon	NA	1636
[C_1C_1Pyrr] [HCOO] [33]	<0.02	514nm-CW- Argon	NA	677
[C_1C_1Pyrr] [CH_3COO] [33]	2.4	514nm-CW- Argon	NA	26362
[PR_4] [xan][28]	NA	910 nm-3ps-76MHz	NA	0.09
[PR_4] [CoCB][28]	NA	910 nm-3ps-76MHz	NA	0.22
[PR_4] [DDTC][26,28]	NA	910 nm-3ps-76MHz	NA	0.45
[PR_4]$_2$[Co(NCS)$_4$][28]	NA	910 nm-3ps-76MHz	NA	1.9
[PR_4]$_2$[DTMN][26,27,28]	NA	910 nm-3ps-76MHz	NA	6.1

IV.2 Nonlinear Absorption

To study nonlinear absorption behavior of ionic liquids, we applied the Z-scan technique; we used an argon laser (514 nm) of variable power, which was focused by an 80 mm focal length lens. The experimental set up scheme used, is shown in Figure 1. The laser beam waist ω_0 at the focus was measured to be 35 μm, a 20 μm and we used an optical path length cell containing the ionic liquid which was translated across the focal region along the axial direction that is the direction of the propagation laser beam. The transmission of the beam was measured through an aperture placed in the far field using a photodetector fed to the digital power meter.

The nonlinear absorption value of some ionic liquids was determined using [38]:

$$T(z) = 1 - \frac{q_0}{2\sqrt{2}} \frac{1}{[1+\frac{z^2}{z_0^2}]} \qquad (13)$$

Where $q_0 = \beta I_0 L_{eff}$ with $|q_0| \ll 1$.

The ionic liquids [C_4mim][TF2N], [C_2mim][TF2N], [C_2mim][BF4], [C_4mim][BF4], [C_2mim][CF3COO], and [C_4mim][CF3COO] were prepared following literature procedures [39], and synthetized in DCN-UGTO Polymers Laboratory, the absorption spectra of the IL are shown at Figure 2, as it is known, they are transparent at visible.

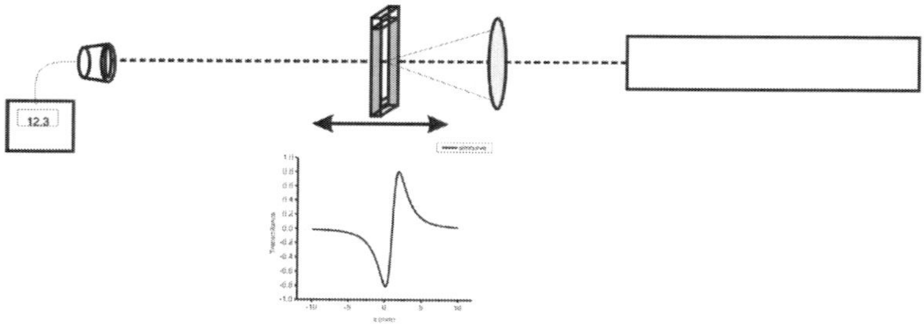

Figure 1. Experimental setup to obtain z-scan curves with open aperture.

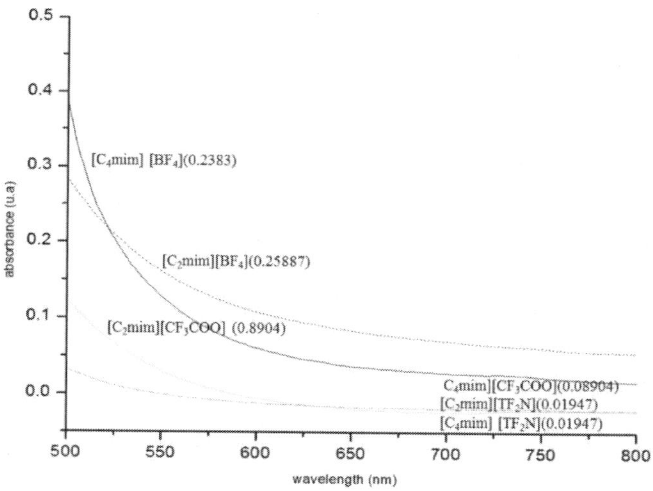

Figure 2. UV-Vis spectra to Ionic liquids.

The refraction curves obtained of ionic liquids, and absorption represented by beta, are shown in Figure 3-5. Figure 3 shows an nonlinear absorption experimental data for [C_2mim][BF4] and [C_4mim][BF4] ionic liquids note the transmittance was decreasing while the IL was approximated to focus where the input intensity was increased, it is the characteristic behavior of Reverse

Saturable Absorption (RSA) or positive nonlinear absorption. [BF4] anion with different cation present the same nonlinear absorption it practically means that the chain length in cation does not influence this property. We measured using a laser at 632.8 nm at 15 mW of power to [C_4mim][BF4] and also presented a RSA behavior. The following ionic liquids [C_2mim CF_3COO] and [C_4mim][CF_3COO] (Figure 4) show a remarkable difference because whilst the first presented a minimum at focus like RSA behavior the second presented a maximum at focus, and this corresponds to RSA-SA switch behavior in the same experimental conditions, which means that the cation has a big influence at nonlinear absorption property. [C_2mim][TF_2N] and [C_4mim][TF_2N] z-scan curves with open aperture are show in Figure 5, they presented an increase at transmittance when ILs are approximate to focus, this means the presence of Saturable Absorption (SA), but note that [C_4mim][TF_2N] will be changed to RSA at higher power than 8 mW.

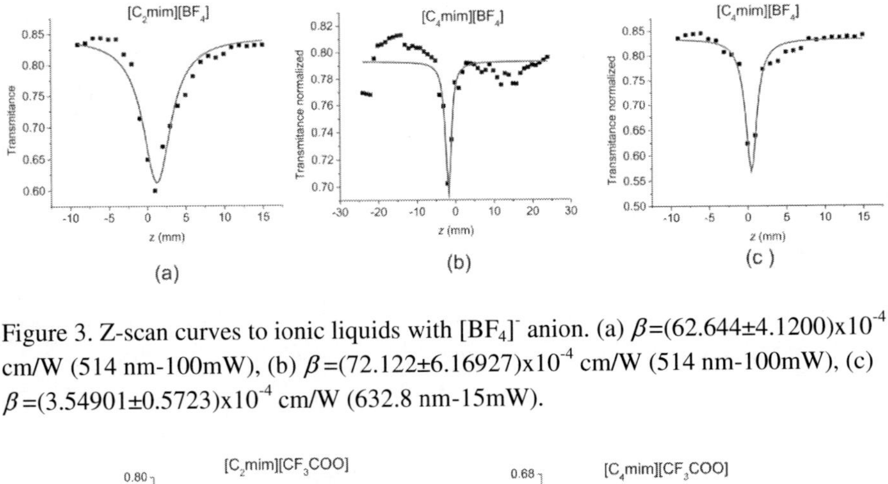

Figure 3. Z-scan curves to ionic liquids with $[BF_4]^-$ anion. (a) β=(62.644±4.1200)x10^{-4} cm/W (514 nm-100mW), (b) β=(72.122±6.16927)x10^{-4} cm/W (514 nm-100mW), (c) β=(3.54901±0.5723)x10^{-4} cm/W (632.8 nm-15mW).

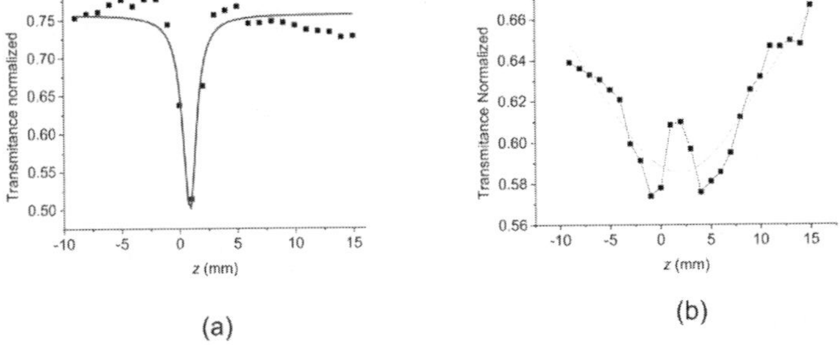

Figure 4. Z-scan curves to ionic liquids with $[CF_3COO]^-$ anion at 514 nm-100mW, (a) β=(70.6339±8.2297)x10^{-4} cm/W, (b) β= NA cm/W.

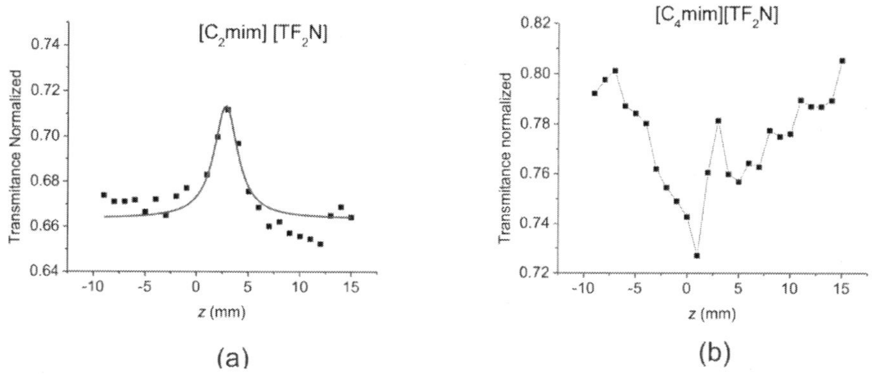

Figure 5. Z-scan curves to ionic liquids with $[TF_2N]^-$ anion at 514 nm-100mW (a) β=(-13.597±1.79465)x10^{-4} cm/W, (b) β = NA cm/W.

The nonlinear optical parameters of presented IL samples, are higher than the reported results under cw regime to dye [35].

V. Nonlinear Optical Properties. Theoretical Calculus

V.1 Dielectric Constants

The dielectric constant (ε) is extracted from the complex permittivity, $\varepsilon^*(\nu)$, measurement of a material as a function of the frequency of the applied electric field. The dielectric constant is defined as the zero-frequency of $\varepsilon^*(\nu)$ in the absence of ionic conductivity and thus reflects the contributions from: a) the ionic, distortion and electric polarization caused by the molecular polarizability (α) and b) the orientation polarization caused by the permanent dipole moment (μ).

Ionic polarization is known to be of importance in crystalline salts below the IR region. Such ion displacements simply lead to conductive motions in molten salts where the ions are small and mobile. In the case of ionic liquids, small relative displacements of the ions, or parts of the ions may be possible but longer-range motions required for conduction hindered. Hence "ionic polarization" may contribute to the overall polarization in this case. Ionic liquids might also display in some cases an additional component of ionic polarization that originates from distortion of the electron density on the ion induced by the presence of surrounding ions and results in a slight displacement of charge. This charge displacement can enhance the overall dipole moment of the ion, resulting in a further contribution to the dielectric constant.. Thus one may expect that ionic polarization may be a contributor to the polarizability of ionic liquids, at some frequency above the region where ion conductions dominates.

The electronic polarizability is only an amount that easily be calculated by using electronic polarizabilities obtained from *ab initio* theory. Table 2 show some of the theoretical dipole moment calculated (μ/D) [40].

Theoretical studies have been used to determine the second hyperpolarizability (γ) of some anions or cations. These were based on computational methods and the calculations was done in the gas phase in absence of any counterions or other lattice//structural interactions present. In general, many studies show that organic and molecular inorganic materials need to have an extensive degree of electronic conjugation, leading to highly polarizable electrons, in order to have a large (γ).

Table 2. Dipole moment calculated to some ions using ab initio

Ion	μ/D
$[BF_4]^-$	0.000
$[PF_6]^-$	0.000
$[Tf_2N]^-$ (*cis*)	3.6902
$[Tf_2N]^-$ (*trans*)	0.299
$[EtSO_4]^-$	5.791
$[Cl]^-$	0.000
$[N(CN)_2]^-$	0.000
$[CF_3COO]^-$	6.679
$[NO_3]^-$	0.000
$[C_2mim]^+$	1.622

Yeates et al. [41] studied the second hyperpolarizability (γ) using *ab initio* Hartree-Fock protocols of some anions as: Trithiocarbonate, xanthatre, Aci-nitro-dithioacetate (K-salt), aci-nitrodiselenoacetate and melonitrile, and some ionic liquids formed by k-salt anion and cations as: Li^+, Na^+,K^+ and MEI^+ . They optimized the geometry of anions using the Restricted Hartree-Fock (RHF) level of theory with Dunning-Hay basis set enhanced with polarization and diffuse *sp* functions. The second hyperpolarizabilities of the anions were computed with finite-field method using the same basis except the difusse sp functions were replaced with the set of diffuse p and d functions designed for NLO computations. They computationally show how there is an enhanced hyperpolarizability from increasing the electron density in the "diffuse" region. And they mention that anions containing higher row elements demonstrate to be a promising expression on this new structure-property relationship. Del Sesto et al. [27] in their chapter calculate the second hyperpolarizability (γ) to some anions using the Restricted Hartree-Fock (RHF) and MP2 protocols (see Table 3). They report the effects of higher row elements which increase the negative charges on γ values of anions and mentioned that the presence of cation due their high charge density can decrease the polarizability of the diffused anions. According to Table 3 , monoanions with third row elements (e.g., $[HOCS_2]^-$) have higher gammas than monoanions with second row elements (e.g,. $[HCO_2]^-$) and dianions containing sulfur ($[O_2NCHCS_2]^{2-}$) have highest gamma values. They mentioned that the presence of cation due their high charge density can decrease the polarizability of the diffuse anions.

Table 3. Theoretical values to some anions using Restricted Hartree-Fock (RHF) and MP2 protocols

Anion	$\gamma x10^{-3}$ a.u (RHF)
$[CS_3]^{2-}$	299[40]
$[OS_3]^{2-}$	50[40]
$[HCS_3]^-$	76[40]
$[HCO_3]^-$	8[40]
$[H_2CS_3]$	20[40]
$[O_2NCHCSe_2]^{2-}$	408[40]
$[NO_2]^-$	9[27]
$[HCO_2]^-$	12[27]
$[HOCS_2]^-$	66[40,27]
$[H_2NCS_2]^-$	84[27]
$[O_2NCHCS_2]^{2-}$	268[40,27]
$[(NCCS)_2]^{2-}$	435[40,27]
Ionic liquid	
$[Li]_2[O_2NCHCS_2]$	42[40]
$[Na]_2[O_2NCHCS_2]$	71[40]
$[K]_2$ $[O_2NCHCS_2]$	119[40]
$[MEI][O_2NCHCS_2]$	66[40]

Table 4. List of names of anions of ILs

Short name anion	Long name anion
$[BF_4]^-$	Tetrafluoroborate
$[PF_6]^-$	Hexafluorophosphate
$[Tf_2N]^-$	Bis(trifluoromethylsulfonyl) imide
$[EtSO_4]^-$	Ethyl Sulfate
$[SbF6]^-$	Hexafluoroantimonate
$[MetSO4]^-$	Methylsulfate
$[MetFSO4^-]$	Triflate
$[Cl]^-$	Chloride
$[N(CN)_2]^-$	Dicyanamide
$[Br]^-$	Bromide
$[CF_3COO]^-$	Trifluoroacetate
$[NO_3]^-$	Nitrate
$[HSO_4]^-$	Hydrogen sulfate
$[HCOO]^-$	Formate
$[CH_3COO]^-$	Acetate
$[DTMN]^-$	Methyl xanthate
$[DDTC]^-$	Dietyldithiacarbamate
$[CoCB]^-$	Bis(dicarbollyl)cobalt(III)
$[Co(NCS)_4]^-$	Tetrathiocyanocobaltate(II)
$[HOCS_2]^-$	Xanthate
$[CS_3]^{2-}$	Trithiocarbonate
$[OS_3]^{2-}$	Yeates
$[HCS_3]^-$	Hydrogen Trithiocarbonate
$[HCO_3]^-$	Hydrogen carbonate
$[H_2CS_3]^-$	Acid Trithiocarbonic
$[O_2NCHCSe_2]^{2-}$	Acid nitrodiselenoacetate
$[NO_2]^-$	Nitrium
$[HCO_2]^-$	Hydrogen carbonaceus
$[N(SO_2CF_3)_2]^-$	N-bis(trifluoromethanesulfonyl)
$[H_2NCS_2]^-$	yeates
$[O_2NCHCS_2]^{2-}$	Aci-nitro-dithioacetate (K-salt)
$[(NCCS)_2]^{2-}$	melonitrile

V.2 Ionic Liquid Crystalline Polymer

Some ionic liquid crystalline polymers, as azobencene ionic liquid crystalline polymer, have been studied. This compound presents nonlinear optical properties and has been studied in solution or thin film, using different techniques as z-scan [42], and OKE and pump-probe techniques [43], nonlinear refraction index value for polymer thin film is n_2 = -1.07×10^{-9} cm^2/W and nonlinear absorption index is $\beta = -0.72 \times 10^4$ cm/GW, were determined using a picosecond laser at 532 nm and z-scan technique also the third order nonlinear susceptibility $\gamma = 8.26 \times 10^{-10}$ esu was determined using OKE and pump-probe techniques, using femtosecond laser pulse generated from a mode-locked Ti:Sapphire oscillator generating 100 fs at 800 nm, with time response as 300 fs; in solution the third order nonlinear susceptibility $\gamma = 2.1 \times 10^{-14}$ esu that corresponds to a second hyperpolarizability $\chi^{(3)} = 2.3 \times 10^{-32}$ esu at concentration of 2×10^{-3} M was determined.

VI. Applications

Modern optical technologies demands improving the ability to control the light, as phase, or intensity to fit, to control, and modify certain new devices. The knowledge of optical properties, such as chromatic dispersion, thermal refraction, linear and nonlinear absorption, and reflectivity, among others, is crucial to design, to manufacture and for the correct operation of a photonic an optoelectronic components. Ionic liquids have a great chance to be used to develop new devices for mainly two reasons: they are considered part of green chemistry which is necessary to contribute to avoid the electronic contamination and because they possess good nonlinear optical properties.

Some Ionics liquids have potential applications as optical limiters, because they exhibit a linear transmission below a threshold and clamps the output to a constant above it. This phenomenon can be achieved by one or more of the nonlinear optical mechanisms such as excited state absorption (ESA), free-carrier absorption (FCA), two photon absorption (TPA), thermal defocusing/scattering, photorefraction, nonlinear refraction, and induced scattering [44].

Ionic liquids are used as plasticizers to develop ion optodes [45], the possibility then exists for using the nonlinear optical properties of ionic liquids to improve optodes or sensors. On the other hand, ionic liquids can be used as solvents to measure thermal nonlinearity of analytes, it has been observed that ILs enhance the sensitivity of measure the thermal nonlinearity and this is due to their relatively better thermooptical properties [46].

Table 5. List of names of cations of ILs

Short name cation	Long name cation
[$C_n$mim] with n=2,4,5,6,8,10	1-C_n-3-methylimidazolium
[C_4C_1Pip]	1-Butyl-1-methylpiperidinium
[C_4Py]	1-Butylpyridinium
[C_1C_1Pyrr]	1-Methyl-1-methylpyrrolidinium
[C_4C_1Pyrr]	1-Butyl-1-methylpyrrolidinium
[C_6C_6 C_6C_{14}P]	Trihexyl(tetradecyl)phosphonium
[C_2C_2 C_2S]	Triethylsulphonium
[PR_4] or [$PC_6C_6C_6C_{14}$]	Trihexyltetradecylphosphonium
[MEI]	1—methyl-3-ethylimidazolium

Acknowledgments

This work was partially supported by Guanajuato University- PIFI 2013, DAIP No. 217/2013 and 223/2013

References

[1] Saadon, HL; Basil, Ali; Adil, A Al-Fregi. "Nonlinear optical properties of new organotellurium compounds containing azomethine and azo groups under CW laser illumination," *Optics & Laser Tech.*, 2014, 58, 33–38.

[2] Mariappan, L; Kandasamy, A; Rathnakumari, M; Sureshkumar, P. "Synthesis, growth and properties of an organic nonlinear optical crystal: 2-Aminopyridinium benzoate (2-APB)," *Optik - International Journal for Light and Electron Optics*, 2013, 124 (17), 2630–2634.

[3] Xinyu, Yang; Weidong, Xiang; Haijun, Zhao; Xiyan, Zhang; Xiaojuan, Liang; Shixun, Dai; Feifei, Chen. "Third-order nonlinear optical

properties of Bi_2S_3 nanocrystals doped in sodium borosilicate glass studied with Z-scan technique," *Mat. Res. Bull.*, 2011, 46 (3), 355–360.

[4] Sabari, TC Girisun; Dhanuskodi, S; Vinitha, G. "$\chi^{(3)}$ measurement and optical limiting properties of metal complexes of thiourea using Z-scan," *Mat. Chem. Phys.*, 2011, 129 (1–2), 9–14.

[5] Non-linear Optics properties of Matter; Edited by: Manthos G. Papadopolus; Andrzej J. Sadlej; Jerzy Leszczynski; Springer, 2003.

[6] Bahaa, EA Saleh; Malvin, Carl Teich. Fundamentals of Photonics, 2nd Edition: Wiley, 2007.

[7] Boyd, RW. Nonlinear optics; Second edition, Amsterdan: Academic Press, 2003; Chap 4.

[8] Saleh, BEA; Teich, MC. Fundamentals of photonics; Wiley Series in Pure and Applied Optics, 1991.

[9] Kovsh, D; Hagan, D; Van Stryland, E. "Numerical modeling of thermal refraction in liquids in the transient regime," *Opt. Exp.*, 2006, 4, 315-327.

[10] Van Stryland, EW; Vanherzeele, H; Woodall, MA; Soileau, MJ; Smirl, AL; Guha, S; Boggess, TF. "Two photon absorption, nonlinear refraction, and optical limiting in semiconductors," *Opt. Eng.*, 1985, 24, 613-623.

[11] Joshi, MP; Swiatkiewicz, J; Xu, F; Prasad, PN; Reinhardt, BA; Kannan, R. "Energy transfer coupling of two-photon absorption and reverse saturable absorption for enhanced optical power limiting," *Opt. Lett.*, 1998, 23, 1742-1744.

[12] Tutt, LW; Boggess, TF. "A review of optical limiting mechanisms and devices using organics, fullerenes, semiconductors and other materials," *Prog. Quant. Electron.*, 1993, 17, 299-338.

[13] Maciek, E Orczyk; Marek, Samoc; Jacek, Swiatkiewicz; Para, N Prasad "Dynamics of Third order nonlinearity of canthaxanthin carotenoid by the optically heterodyned phase tuned femtosecond optical Kerr gate," *J. Chem. Phys.*, 1993, 98, 2524.

[14] Boon, K Teo; Yi, Hui Xu; Bing, Yuan Zhong; Yuan, Kang He; et al. "A comparative study of third –order nonlinear optical properties of silver phenylacetylide and related compounds via ultrafast optical Kerr effect measurements," *Inorg. Chem.*, 2001, 40, 6794-6801.

[15] Sheik-Bahae M; Said AA; Wei TH; Hagan DJ; Van-Stryland, EW. "Sensitive measurement of optical nonlinearities using a single beam," *IEEE J. Quant. Electron*, 1990, 26, 760.

[16] Carter, CA; Harris, JM. "Comparison of models describing the thermal lens effect", *Appl. Opt.*, 1984, 23, 476-481 ()

[17] Gordon, JP; Leite, RCC; Moore, RS; Porto, SPS; Whinnery, JR. "Long transient effects in laser with Inserted liquid samples", *J. Appl. Phys.*, 1965, 36(1), 3-8.

[18] Fabio Luiz Sant'Anna Cuppo; Antonio Martins Fihuereido Neto; Sergio Leonardo Gómez, "Thermal-lens model compared with the Sheik Bahae formalism in interpreting Z-scan experiments on lyotropic liquid crystals" *J. Opt. Am.B*, 2002, 19(6), 1342.

[19] Garcia, EV Ramirez; Arroyo, ML Carrasco; Mendez, MM Otero; Reynoso, E Lara; Chavez, S Cerda; Iturbe, MD Castillo. "Z-scan and spatial self-phase modulation of a Gaussian beam in a thin nonlocal nonlinear media," *J. Opt.*, 2011, 13, 085203.

[20] Rashidian, MR Vaziri; Hajiesmaeilbaigi, F; Maleki, MH. "Generalizing the Z-scan Theory for nonlocal nonlinear media," *J. Opt.*, 2013, 15, 025201.

[21] Garcia, EV Ramirez; Arroyo, ML Carrasco; Mendez, MM Otero; Chavez, S Cerda; Iturbe, MD Castillo. " Far field intensity distributions due to spatial self phase modulation of a Gaussian beam by a thin nonlocal nonlinear media," *Opt. Exp.*, 2010, 18, 22067-22079.

[22] Bing, Gu; Jing, Chen; Ya, Xian, Fan; Jianping, Ding; Hui-Tian, Wang. "Theory of Gaussian beam Z scan with simultaneous third- and fifth-order nonlinear refraction based on a Gaussian decomposition method," *J. Opt. Soc. Am. B*, 2005, 22, 2651-2659.

[23] Sheldon, SJ; Knight, LV; Thorne, JM. "Laser-induced thermal lens effect: a new theoretical model," *Appl. Opt.*, 1982, 21, 9, 1663.

[24] Severiano, I; Méndez, MM; Arroyo, ML; Iturbe, MD. "Comparison of different models employed to describe the z-scan curves for thick nonlinear optical media," *J. Mod. Opt.*, 2013, 60(3), 248-254.

[25] Souza, RF; Alencar, M; Meneghetti, MR; Dupont, J; Hickmann, JM. "Nonlocal optical nonlinearity of ionic liquids," *J. Phys. Condens. Matter.*, 2008, 20, 155102 .

[26] Del Sesto RE; Doug, S Dudis; Fassil, Ghebremichael; Norman, E Heimer; Tammy, KC Low; John, S Wilkes; Todd, A Yeates. "Modeling, synthesis and characterization of third-order nonlinear optical salts," *Proc. SPIE*, 2003, 5212, 292-298.

[27] Del Sesto, RE; Cynthia, Corley; Robertson, Al; John, S Wilkes, "Tetraalkylphosphonium-based ionic liquids," *Journal of Organometallic Chemistry*, 2005, 690, 2536–2542.

[28] Del Sesto, RE; Doug, S Dudis; Fassil, Ghebre Michael; Norman, E Heimer; Tammy, KC Low; John, S Wilkes; Todd, A Yeates. Nonlinear Optical Ionic Liquids, Chap. 12 In Ionic Liquids IIIB: Fundamentals, Progress, Challenges, and Opportunities; Rogers R., Et al; ACS Symposium Series; American Chemical Society; Washington D.C., 2005.

[29] Cassio, EA Santos; Marcio, ARC Alencar; Pedro, Migowski; Jairton, Dupont; Jandir, M Hickmann. "Anionic and cationic influence on the nonlocal nonlinear optical response of ionic liquids," *Chem. Phys.*, 2012, 403, 33-36.

[30] Novoa-Lopez, JA; Lopez, E Lago; Dominguez-Perez, M; Troncos, J; Varela, LM; de la R Fuente; Cabeza, O; Michinel, H; Rodriguez, JR. "Thermal refraction in ionic liquids induced by a train of femtosecond laser pulses,"*Optics & Laser Tech.*, 2014, 61, 1-7.

[31] Chieu, D Tran; Santhosh, Challa; Mladen, Franko. "Ionic Liquids as an Attractive Alternative Solvent for Thermal Lens Measurements," *Anal. Chem..*, 2005, 77, 7442-7447.

[32] Cassio, EA Santos; Marcio, ARC Alencar; Pedro, Migowski; Jairton, Dupont; Jandir, M Hickmann. "Nonlocal nonlinear optical response ionic liquids under violet excitation," Hindawi Publishing Corporation, *Adv. Matt. Sci. Eng.*, 2013, Article ID 104914, http://dx.doi.org/10.1155/2013/104914

[33] Valencia-Loredo, CE; Barrera-, Rivera, KA; Trejo-Durán, M; Alvarado-Méndez, E; Martínez, Richa, A; Andrade-Lucio, JA. "Nonlinear optical characterization of ionic liquids, nonlinear optical characterization of ionics liquids," *Proc. SPIE,* 7386. Photonics North;, 2009, 738610 doi: 10.1117/12.839505

[34] Franco, M; Tran, CD. "Thermal lens effect in Electrolyte and surfactant media," *J. Phys. Chem.*, 1991, 95, 6688-6696.

[35] Pramodini, S; Poornesh, P; Nagaraja, KK; et al. "Thermally induced nonlinear optical response and optical power limiting of acid blue 40 dye," *Curr. Appl. Phys.*, 2013, 13, 1175-1182.

[36] Dong, Xiao; Larry, G Hines Jr; Mark, W Holtz; Kihyung, Song; Richard, A Bartsch; Edward, L Quitevis. "Effect of cation symmetry on the low-frequency spectra of imidazolium ionic liquids: OKE and Raman spectroscopic measurement and DFT calculations, " *Chem. Phys. Lett.*, 2010, 497, 37-42

[37] Trejo-Durán, M; Alvarado-Méndez, E; Barrera-Rivera, KA; Castaño, VM; "Nonlinear Optical phenomena in Bi-ionic Liquids Bi-ionic liquids" send to *JEOS* (2014).

[38] W Van Stryland, E; Sheik-Bahae, M; "Characterization Techniques and Tabulations for Organic Nonlinear Materials," M. G. Kuzyk and C. W. Dirk, Eds., page 655-692, Marcel Dekker, Inc.,, 1998

[39] Aniruddha, P; Prasun-Kumar, M; Anunay, S "On the Optical Properties of the Imidazolium Ionic Liquids", *J. Phys. Chem. B*, 2005, 109 (18), 9148-9153.

[40] Izgorodina, E; Forsyth, M; R MacFarlane, D; "In the components of the dielectric constants of ionic liquids: ionic polarization?," *Phys. Chem. Chem. Phys.*, 2009, 11, 2452-2458.

[41] Yeates, AT; Dudis, DS; Wilkes, JS. "Nonlinear Optical Properties of the Salts of Diffuse anions". *Proc. SPIE-Int. Soc. Opt. Eng.*, 2000, 4106, 334.

[42] Xiaoqiang, ZXhang; Changshun, Wang; Xu, Pan; Sufang, Xiao; Tingchao, He; Xuemin, Lu. "Nonlinear optical properties and photoinduced anisotropy of an azobenzene liquid crystalline polymer," *Opt. Comm.*, 2010, 283, 146-150.

[43] Fuli, Zhao; Changshun, Wang; Yi, Zeng; Zuanming, Jin; Guohong, Ma, "Ultrafast third-order nonlinear optical properties of an azobenzene-containing ionic liquid crystalline polymer," *Chem. Phys. Lett.*, 2013, 558, 100-103.

[44] Nagaraja, KK; Pramodini, S; Santhoshkumar, A; Nagaraja, HS; Poornesh, P; Kekuda, D. "Third-order nonlinear optical properties of Mn doped ZnO thin films under CW laser illumination," *Opt. Mater.*, 2013, 35, 431-439.

[45] Jingwei, Zhu; Jingying, Zhai; Xue, Li; Yu, Qin. "Applications of hydrophobic room temperature ionic liquids in ion-selective optodes," *Sensors and Actuators B*, 2011, 159, 256–260.

[46] Chieu, D Tran; Santhosh, Challa; Mladen, Franko. "Ionic Liquids as an Attractive Alternative solvent for Thermal Lens Measurements," *Anal. Chem.*, 2005, 77, 7442-7447.

INDEX

A

B

E

F

G

H

I

J

K

L

M

N

O

P

Q

R

S

T

U

V

W

Y

Z